Flat Spacetime Gravity (FSG)

An Alternative Theory of Gravitation

THANASIS EFTHYMIOU

ISBN (Hardcover): 978-618-88347-3-6
ISBN (Paperback): 978-618-88347-2-9

Athens, 2026

To the great figures of the past

Contents

1. Introduction

Gravitation remains one of the deepest and most conceptually demanding subjects of theoretical physics. Since the formulation of general relativity, gravitational phenomena have been described through the geometry of curved spacetime itself, encoded in a dynamical metric governed by Einstein's field equations. This geometric paradigm has achieved remarkable experimental success, ranging from precision solar– system tests to gravitational–wave observations and cosmological expansion.

Nevertheless, the geometric formulation of gravity is not the only logically consistent framework within which relativistic gravitational phenomena may be described. An alternative viewpoint treats gravitation not as curvature of spacetime, but as a relativistic field dynamical interaction evolving on a fixed flat Minkowski background. Historically, flat–spacetime approaches have often encountered difficulties in preserving full covariance or reproducing relativistic effects without implicitly reintroducing geometric structures.

The present work develops a coherent realization of this flat–spacetime programme. A complete covariant field–theoretic description of gravitation is constructed using antisymmetric field tensors, observer–based projections, and a unified Lagrangian–Hamiltonian formalism. No curvature of spacetime is postulated at any stage. All gravitational phenomena arise solely from interactions mediated by the gravitational field bivectors and their associated scalar and vector potentials, defined and evolving entirely within Minkowski spacetime.

A central theoretical component of the formulation is the decomposition of antisymmetric tensors into orbital and Fermi–type sectors relative to a chosen four–velocity. This geometrically transparent procedure provides a natural separation between spatial dynamics and mixed time–space coupling, allowing both classical motion and relativistic corrections to be described consistently within flat spacetime kinematics. Within this framework, gravitation assumes a structural form closely analogous to Maxwellian field theory — not merely as a formal analogy, but as a genuine dynamical correspondence expressed through parallel field equations and conservation laws, while remaining fully covariant.

On this basis the theory is successively developed through a Lagrangian formulation, a canonical Hamiltonian description, and a semiclassical WKB analysis leading to a de Broglie matter–wave interpretation. Classical trajectories, wave propagation, and quantum–like dynamics emerge as coherent aspects of a single unifying field framework rather than as independent postulates.

The observational relevance of the theory is investigated through direct calculation of the principal gravitational tests. Gravitational redshift results from local time–rate variations induced by the scalar potential; orbital precession and frame effects arise from velocity–dependent couplings; and the deflection of light follows naturally from orbital interactions alone, without invoking spacetime curvature. At cosmological scales, the evolution of the gravitational time–rate factor $\gamma(t, r)$ yields effective expansion, redshift, and density–evolution phenomena consistent with observations, likewise within a flat–spacetime setting.

The purpose of this monograph is therefore not to challenge the empirical achievements of general relativity, but to demonstrate that a wide class of gravitational phenomena conventionally attributed to spacetime curvature can equally be described within a fully covariant flat–spacetime field theory. The resulting framework of Flat Spacetime Gravity (FSG) provides a unified mathematical structure in which classical dynamics, wave behavior, quantum semiclassical effects, and cosmological evolution coexist as different manifestations of a single relativistic field system, inviting further analytical development and phenomenological investigation. The chapters that follow develop this program in detail. Section 2 establishes the kinematical and geometric foundations of the theory. Sections 3–5 construct the gravitational field equations and their classical observational consequences. Sections 6 and 7 introduce the Lagrangian and Hamiltonian structure. Sections 8 through 10 examine semiclassical and quantum aspects through de Broglie and WKB formulations. Finally, Section 11 presents the cosmological sector of the theory, culminating in a flat– spacetime description of large–scale gravitational dynamics.

2. Kinematic Framework

2.1 Introductory Remark

In this section we construct the purely kinematical sector of the Flat Spacetime Gravity (FSG). Its structure mirrors the electromagnetic decomposition $(\phi, \mathbf{A}) \rightarrow (\mathbf{E}, \mathbf{B})$, but is applied to the four–velocity field U^{μ} itself through a *kinematic potential*. All relativistic inertial effects — acceleration, vorticity, and the relativistic hydrodynamic Euler equation — arise from this construction alone, without invoking any gravitational field.

2.2 Four–Velocity and 3+1 Decomposition

The four–velocity of a timelike worldline is

$$U^{\mu} = \gamma(c, \mathbf{v})$$
$$U_{\mu} = \gamma(-c, \mathbf{v})$$
$$\gamma = \frac{1}{\sqrt{1 - v^2/c^2}}$$

thus, the spatial proper–velocity is

$$\mathbf{u} = \gamma\,\mathbf{v}$$

The four–current mass density is

$$\mathrm{J}^{\mu} = \rho_0\, \mathrm{U}^{\mu} = (\rho_0 \gamma c, \mathbf{j}) = (\rho_0 \gamma c, \rho_0 \mathbf{u})$$
$$\mathbf{j} = \rho_0 \mathbf{u} = \rho_0 \gamma \mathbf{v}$$

The four–acceleration is

$$\alpha^{\mu} = \frac{d\mathrm{U}^{\mu}}{d\tau} = (\alpha^0, \mathbf{a})$$
$$\mathbf{a} = \frac{d\mathbf{u}}{d\tau}$$

2.3 Kinematic Potential $\Phi_{\mu}^{(\mathrm{kin})}$

We define a four–potential built directly from the four–velocity:

$$\Phi^{(\mathrm{kin})\,\mu} = \left(\frac{\phi^{(\mathrm{kin})}}{c},\ A^{(\mathrm{kin})\,\mu}\right)$$

$$\Phi_{\mu}^{(\mathrm{kin})} = \left(-\frac{\phi^{(\mathrm{kin})}}{c},\ A_{\mu}^{(\mathrm{kin})}\right)$$

The key identification is

$$\Phi^{(\mathrm{kin})\,\mu} = -U^{\mu}$$

$$\Phi_{\mu}^{(\mathrm{kin})} = -U_{\mu}$$

In components we choose

$$\phi^{(\mathrm{kin})} = -\gamma c$$

$$\mathbf{A}^{(\mathrm{kin})} = -\mathbf{u}$$

so that the temporal and spatial parts of $\Phi_{\mu}^{(\mathrm{kin})}$ encode directly the Lorentz factor and the proper–velocity.

2.4 Kinematic Fields

In analogy with electromagnetism, we define

$$\boldsymbol{\delta} = -\nabla\phi^{(\mathrm{kin})} - \frac{\partial \mathbf{A}^{(\mathrm{kin})}}{\partial t}$$

$$\boldsymbol{\xi} = \nabla \times \mathbf{A}^{(\mathrm{kin})}$$

Substituting $\phi^{(\mathrm{kin})} = -\gamma c$ and $\mathbf{A}^{(\mathrm{kin})} = -\mathbf{u}$ we obtain

$$\boldsymbol{\delta} = \mathrm{c}^2\,\nabla\gamma + \frac{\partial \mathbf{u}}{\partial \mathrm{t}}$$

$$\boldsymbol{\xi} = -\nabla \times \mathbf{u}$$

Thus

- $\boldsymbol{\delta}$ encodes inertial changes in γ and $\mathbf{u}$ (the "electric–like" inertial part),
- $\boldsymbol{\xi}$ encodes the vorticity of the congruence (the "magnetic–like" inertial part).

2.5 Kinematic Tensor $W_{\mu\nu}$

The antisymmetric kinematic tensor is defined by

$$W_{\mu\nu} = \partial_\mu \Phi_\nu^{(\mathrm{kin})} - \partial_\nu \Phi_\mu^{(\mathrm{kin})} = -(\partial_\mu \mathrm{U}_\nu - \partial_\nu \mathrm{U}_\mu) = \partial_\nu \mathrm{U}_\mu - \partial_\mu \mathrm{U}_\nu$$

Using the decomposition above one finds

$$W_{0i} = -\, c\, \delta_i$$
$$W_{ij} = \epsilon_{ijk}\, \xi_k$$

Hence the full 4×4 matrix is

$$W_{\mu\nu} = \begin{pmatrix} 0 & -\frac{\delta_1}{c} & -\frac{\delta_2}{c} & -\frac{\delta_3}{c} \\ \frac{\delta_1}{c} & 0 & \xi_3 & -\xi_2 \\ \frac{\delta_2}{c} & -\xi_3 & 0 & \xi_1 \\ \frac{\delta_3}{c} & \xi_2 & -\xi_1 & 0 \end{pmatrix}$$

Note that, in accordance with the sign convention adopted in the theory (the spatial 3x3 block equals - $\boldsymbol{\xi}$), the component satisfies W_{13}

$$W_{13} = -\,\xi_2$$

2.6 Acceleration

2.6.1 Acceleration as $a_\mu = W_{\mu\nu}U^\nu$

Let a_μ be the covariant four–acceleration,

$$a_\mu = \frac{dU_\mu}{d\tau} = U^\nu\,\partial_\nu U_\mu$$

From the definition of $W_{\mu\nu}$ and $\Phi_\mu^{(\mathrm{kin})} = -U_\mu$ we have

$$W_{\mu\nu} = \partial_\mu \Phi_\nu^{(\mathrm{kin})} - \partial_\nu \Phi_\mu^{(\mathrm{kin})} = -\,\partial_\mu U_\nu + \partial_\nu U_\mu$$

Then

$$W_{\mu\nu}U^\nu = -\,(\partial_\mu U_\nu)U^\nu + (\partial_\nu U_\mu)U^\nu$$

Now $U_\nu U^\nu = -c^2$ is constant, hence

$$\partial_\mu(U_\nu U^\nu) = 2(\partial_\mu U_\nu)U^\nu = 0$$
$$\Rightarrow\ (\partial_\mu U_\nu)U^\nu = 0$$

Therefore

$$W_{\mu\nu}U^\nu = (\partial_\nu U_\mu)U^\nu = U^\nu\,\partial_\nu U_\mu = \frac{dU_\mu}{d\tau} = a_\mu$$

Thus, we have established the purely kinematical identity

$$a_\mu = W_{\mu\nu}U^\nu$$

which will later be promoted to the dynamical law once gravitational fields are included.

For the spatial components we obtain

$$\mathbf{a} = \gamma\,\boldsymbol{\delta} + \mathbf{u} \times \boldsymbol{\xi}$$

2.6.2 Spatial Form and Relativistic Euler Equation

Substituting in

$$\mathbf{a} = \gamma\, \boldsymbol{\delta} + \mathbf{u} \times \boldsymbol{\xi}$$

the explicit expressions for $\boldsymbol{\delta}$ and $\boldsymbol{\xi}$ we obtain

$$\mathbf{a} = \gamma\left(c^2\nabla\gamma + \frac{\partial \mathbf{u}}{\partial t}\right) - \mathbf{u} \times (\nabla \times \mathbf{u})$$

Using

$$U^\mu U_\mu = -c^2, \qquad u^2 = c^2(\gamma^2 - 1), \qquad \gamma\nabla\gamma = 1/2\,\nabla(\gamma^2)$$

$$\mathbf{a} = \nabla(1/2\,u^2) + \gamma\,\frac{\partial \mathbf{u}}{\partial t} - \mathbf{u} \times (\nabla \times \mathbf{u})$$

This is the *fully relativistic hydrodynamic Euler equation* arising purely from kinematics (no forces, no gravity):

$$\mathbf{a} \;=\; \nabla(1/2\,u^2) - \mathbf{u} \times (\nabla \times \mathbf{u}) + \gamma\,\frac{\partial \mathbf{u}}{\partial t}$$

In stationary flow ($\partial \mathbf{u}/\,\partial t = 0$) this reduces to

$$\mathbf{a} = \nabla(1/2\,u^2) - \mathbf{u} \times (\nabla \times \mathbf{u})$$

which is precisely the relativistic generalization of the Euler equation for an ideal fluid.

2.7 Angular Momentum Bivector $H_{\mu\nu}$

The specific (per unit mass) angular momentum is encoded in the bivector

$$H_{\mu\nu} \;=\; X_\mu U_\nu - X_\nu U_\mu \;=\; (X \wedge U)_{\mu\nu}$$

where $X^\mu = (ct, \mathbf{r})$ is the position four–vector.

In a given inertial frame, the spatial dual of $H_{\mu\nu}$ is

$$\mathbf{h} = \mathbf{r} \times \mathbf{u}$$

so that

$$H_{ij} = \epsilon_{ijk}\, h_k$$
$$H_{0i} = -c\,(\gamma\mathbf{r} - \mathbf{u}t)_i$$

The mixed components H_{0i} do not correspond to a spatial observable; they encode the relativity of simultaneity between the particle's time coordinate and its spatial angular momentum.

For a central inertial motion (no external torque) one has

$$\frac{dH_{\mu\nu}}{d\tau} = 0 \quad \Rightarrow \quad \frac{d\mathbf{h}}{d\tau} = 0$$

that is, the specific angular momentum **h** is conserved. This kinematic fact will later justify the use of a constant

$$h = |\mathbf{h}| = r^2\,\frac{d\theta}{d\tau}$$

in the analysis of orbital motion and perihelion precession.

2.8 The Spatial Projector $P(V)$

2.8.1 The general form of the spatial projector

For any unit timelike four–vector V^μ we define the spatial projector onto the three–dimensional hyperplane orthogonal to V^μ by

$$P(V)_{\mu\nu} = \eta_{\mu\nu} + \frac{V_\mu V_\nu}{c^2}$$

This tensor satisfies

$$P(V)_{\mu\nu}V^\nu = 0$$
$$P(V)_\mu^\alpha\, P(V)_\alpha^\nu = P(V)_\mu^\nu$$

and therefore. acts as the unique covariant map that removes all components parallel to V^μ, leaving only the spatial content as measured in the instantaneous rest–frame determined by V^{μ}. Given any tensor $A_{\mu\nu}$, the projection of this tensor

$$A_{\mu\nu}^{\perp(V)} = P(V)^{\alpha}_{\mu} A_{\alpha\beta} P(V)^{\beta}_{\nu}$$

represents precisely the part of $A_{\mu\nu}$ that belongs to the three–dimensional rest–space orthogonal to V^{μ}.

2.8.2 Application for the test body $V = U$

In the kinematical decomposition of the tensor $A_{\mu\nu}$ associated with the worldline of the test body, the relevant four–vector is the particle's own four–velocity U^{μ}. Evaluating the spatial projector at $V^{\mu} = U^{\mu}$ gives

$$P(U)_{\mu\nu} = \eta_{\mu\nu} + \frac{U_{\mu}U_{\nu}}{c^2}$$

and the purely orbital (spatial) part of the tensor $A_{\mu\nu}$ is defined by

$$A_{\text{particle},\mu\nu} = P(V)^{\alpha}_{\mu} A_{\mu\nu} P(V)^{\beta}_{\nu}$$

which represents the purely spatial content of the tensor $A_{\mu\nu}$ as measured in the *particle's* rest–frame $V = U = (c, 0,0,0)$.

In the instantaneous rest frame of the particle, where $U^{\mu} = (c, 0,0,0)$, the 3×3 spatial block of $A_{\text{particle},\mu\nu}$ coincides with the spatial block of $A_{\mu\nu}$. In a general laboratory frame, however, they differ by a specific term, reflecting the fact that $P(U)$ removes all components parallel to U^{μ}.

2.8.3 Application for the source $V = N$

When analyzing the interaction between the test body and a gravitational source, the relevant rest–frame is that of the source itself. Let N^{μ} denote the four–velocity of the source. The corresponding projector is

$$P(N)_{\mu\nu} = \eta_{\mu\nu} + \frac{N_\mu N_\nu}{c^2}$$

The projection of the tensor $A_{\mu\nu}$ of the test body, evaluated as observed in the source's rest–frame, is then defined by

$$A_{\text{source},\mu\nu} = P(N)^{\alpha}_{\mu}\, A_{\mu\nu}\, P(N)^{\beta}_{\nu}$$

which represents the purely spatial content of the tensor $A_{\mu\nu}$ as measured in the *source's* rest–frame $V = N = (c, 0,0,0)$.

2.9 Decomposing the Kinematical Tensor $W_{\mu\nu}$

2.9.1 Introductory Remark

The decomposition of the kinematic tensor $W_{\mu\nu}$ into an orbital part and a Fermi–Walker part is not merely an algebraic rearrangement. It reflects a deep geometric fact: a single antisymmetric tensor can describe, at once, both the intrinsic non–rotating evolution along a worldline and the projected orbital rotation seen by an external observer. What distinguishes these two contributions is the role of the spatial projector $P_{\mu\nu} = \eta_{\mu\nu} + U_\mu U_\nu / c^2$, which extracts the laboratory–measured rotational content from $W_{\mu\nu}$, leaving behind a Fermi–Walker remainder that governs the evolution of all comoving geometric objects. In the following subsection we build this split carefully, show why it is unique, and demonstrate how it controls the dynamics of vectors and tensors along the worldline.

2.9.2 The basic 4D tensors

The starting point is the antisymmetric "universal'' tensor $W_{\mu\nu}$,

$$W_{\mu\nu} = -\left(\partial_\mu U_\nu - \partial_\nu U_\mu\right) = \partial_\nu U_\mu - \partial_\mu U_\nu$$

In flat–spacetime kinematics the fundamental antisymmetric tensor $W_{\mu\nu}$ is

$$a^\mu = W_\lambda^{\ \mu} U^\lambda$$
$$W_{\mu\nu} = -W_{\nu\mu}$$

where U^μ is the four–velocity and $a^\mu = dU^\mu/d\tau$ the four–acceleration.

It is natural to ask whether $W_{\mu\nu}$ admits a meaningful internal decomposition into physically distinct rotational contributions. As we show below, such a decomposition indeed exists and reflects a clean splitting into *Fermi* and *Orbital* parts. This splitting is fully covariant, frame–independent in its formulation, and captures the exact kinematic structure measured by the laboratory observer.

Our aim is to decompose $W_{\mu\nu}$ into two geometrically distinct parts,

$$W_{\mu\nu} = \Omega_{\mu\nu}^{(\mathrm{orb})} + \Omega_{\mu\nu}^{(\mathrm{F})}$$

where:

- $\Omega_{\mu\nu}^{(\mathrm{F})}$ is the Fermi-Walker (Thomas–type) piece, directly built from U^μ and a^μ.
- $\Omega_{\mu\nu}^{(\mathrm{orb})}$ is the "orbital" part: it contains the residual vorticity of the worldline once the Fermi contribution has been subtracted.

2.9.3 The Fermi-Walker Tensor

Independently, the Fermi-Walker tensor is defined by

$$\Omega^{(\mathrm{F})} = \frac{1}{c^2}\, \mathrm{U} \wedge \alpha = \frac{1}{c^2}(U\, a^T - a\, U^T)\Omega_{\mu\nu}^{(\mathrm{F})} = \frac{1}{c^2}(U_\mu a_\nu - U_\nu a_\mu)$$
$$\Omega_{\mu\nu}^{(\mathrm{F})} = -\Omega_{\nu\mu}^{(\mathrm{F})}$$

A direct contraction with U^ν gives

$$\Omega_{\mu\nu}^{(\mathrm{F})} U^\nu = \frac{1}{c^2}\left(U_\mu\, a_\nu U^\nu - U_\nu U^\nu\, a_\mu\right)$$

Since $U_\nu U^\nu = -c^2$ and $a_\nu U^\nu = 0$, we obtain

$$\Omega^{(F)}_{\mu\nu}\, U^{\nu} = a_{\mu}$$

Thus $\Omega^{(F)}_{\mu\nu}$ is precisely the unique antisymmetric tensor whose action on U^{ν} reproduces the 4–acceleration:

$$a^{\mu} = \Omega^{(F)\mu}_{\nu} U^{\nu}$$

To expose the internal structure behind $\Omega^{(F)}_{\mu\nu}$, we introduce the symmetric tensor

$$Y \equiv \mathrm{U}\,\mathrm{U}^{\mathrm{T}}$$

$$Y_{\mu\nu} \equiv U_{\mu}\, U_{\nu}$$

As will be shown in later sections, the standard stress–energy tensor for a dust fluid is

$$M_{\mu\nu} = \rho_0\, U_{\mu} U_{\nu}$$

Hence $Y_{\mu\nu}$ is simply the *unit–mass stress–energy tensor*:

$$Y_{\mu\nu} = \frac{1}{\rho_0} M_{\mu\nu}$$

The role of $Y_{\mu\nu}$ is to extract the component of $W_{\mu\nu}$ that is in line with the matter current density of the test particle.

Using $a_{\mu} = W^{\lambda}_{\mu}\, U_{\lambda}$ in the definition of $\Omega^{(F)}_{\mu\nu}$, gives

$$\Omega^{(F)}_{\mu\nu} = \frac{1}{c^2}\left(W^{\lambda}_{\nu}\, U_{\lambda} U_{\mu} - W^{\lambda}_{\mu}\, U_{\lambda} U_{\nu}\right)$$

For the first term we use antisymmetry:

$$W^{\lambda}_{\nu}\, U_{\lambda}\, U_{\mu} = -\, U_{\mu} U_{\lambda}\, W^{\lambda}_{\nu} = -\,(YW)_{\mu\nu}$$

The second term is immediately

$$W^{\lambda}_{\mu} U_{\lambda} U_{\nu} = (WY)_{\mu\nu}$$

$$(WY)_{\mu\nu} = W_{\mu}^{\lambda} Y_{\lambda\nu}$$

Therefore,

$$\Omega_{\mu\nu}^{(F)} = -\frac{1}{c^2}\left[(WY)_{\mu\nu} + (YW)_{\mu\nu}\right]$$

i.e.

$$\Omega^{(F)} = -\frac{1}{c^2}(\mathrm{WY} + \mathrm{YW})$$

Using $M_{\mu\nu} = \rho_0 Y_{\mu\nu}$, this can be written equivalently as

$$\Omega^{(F)} = -\frac{1}{\rho_0 c^2}(\mathrm{WM} + \mathrm{MW})$$

This expression shows that the Fermi-Walker tensor is not an independent geometric object. It is generated directly from the kinematical tensor $W_{\mu\nu}$ through a *symmetrized coupling* with the *unit–mass stress–energy tensor* $Y_{\mu\nu} = U_\mu U_\nu$ (or, equivalently, with the full stress–energy tensor $M_{\mu\nu}$). Thus $\Omega^{(F)}$ represents precisely the portion of the kinematical structure $W_{\mu\nu}$ that is "filtered" by the energy–momentum content flowing along the worldline U^μ.

2.9.4 The Orbital Tensor

The orbital part is then defined by the residual

$$\Omega_{\mu\nu}^{(orb)} = W_{\mu\nu} - \Omega_{\mu\nu}^{(F)}$$
$$\Omega_{\mu\nu}^{(orb)} = -\Omega_{\nu\mu}^{(orb)}$$

By construction,

$$\Omega_{\mu\nu}^{(orb)} U^\nu = \left(W_{\mu\nu} - \Omega_{\mu\nu}^{(F)}\right) U^\nu = a_\mu - a_\mu = 0$$

$$\Omega_{\mu\nu}^{(orb)} U^\nu = 0$$

Hence $\Omega_{\mu\nu}^{(\mathrm{orb})}$ is purely spatial in the local rest–space of U^{μ}: it encodes the "orbital" vorticity of the worldline, while $\Omega_{\mu\nu}^{(\mathrm{F})}$ is responsible for the actual 4–acceleration.

Having defined

$$W_{\mu\nu} = -(\partial_{\mu}U_{\nu} - \partial_{\nu}U_{\mu})$$
$$\Omega_{\mu\nu}^{(\mathrm{F})} = \frac{1}{c^2}\,(U_{\mu}a_{\nu} - U_{\nu}a_{\mu})$$
$$\Omega_{\mu\nu}^{(\mathrm{orb})} = W_{\mu\nu} - \Omega_{\mu\nu}^{(\mathrm{F})}$$

we now apply the standard orthogonal projector

$$P(V)_{\mu\nu} = \eta_{\mu\nu} + \frac{V_{\mu}V_{\nu}}{c^2}$$

onto the rest–space of the particle, so $V^{\mu} = U^{\mu}$:

$$P_{\mu\nu} = \eta_{\mu\nu} + \frac{1}{c^2}U_{\mu}U_{\nu}$$
$$P_{\mu\nu}\,U^{\nu} = 0$$
$$P_{\mu}^{\alpha}\,P_{\alpha\nu} = P_{\mu\nu}$$

A key identity, as we have seen in the previous subsections, is that the purely spatial part of an antisymmetric tensor $A_{\mu\nu}$ in the rest–space of U^{μ} is obtained via the double projection

$$A_{\mu\nu}^{\perp} = P_{\mu}^{\alpha}\,A_{\alpha\beta}\,P_{\nu}^{\beta}$$

Applying this to $W_{\mu\nu}$, we obtain the *orbital* part:

$$\Omega_{\mu\nu}^{(\mathrm{orb})} = P_{\mu}^{\alpha}\,W_{\alpha\beta}\,P_{\nu}^{\beta}$$

$$\Omega^{(orb)} = P\,\mathrm{W}\,\mathrm{P}$$

The tensor $\Omega^{(orb)} = PWP$ represents the projection of the full kinematical tensor $W_{\mu\nu}$ onto the three–dimensional rest–space orthogonal to the four–velocity U^{μ}. This projected tensor encodes precisely the part of $W_{\mu\nu}$ that is purely spatial for an observer comoving with the particle. In the *instantaneous rest frame* of the particle, where $U^\mu = (c, 0,0,0)$ and consequently $P_{\mu\nu} = \mathrm{diag}(0,1,1,1)$, the 3×3 spatial block of $\Omega^{(orb)}$ coincides exactly with the spatial block of $W_{\mu\nu}$. In a general laboratory frame, however, the spatial components differ because the projection removes the Fermi–Walker contribution $\Omega_{\mu\nu}^{(\mathrm{F})} = 1/c^2\,(U_\mu a_\nu - U_\nu a_\mu)$, causing $\Omega_{ij}^{(\mathrm{orb})} \neq W_{ij}$ unless evaluated in the instantaneous rest frame of the particle. Thus, $\Omega^{(orb)}$ should be interpreted as the *particle's rest–frame spatial content* of W, not merely its spatial block in an arbitrary coordinate system.

Indeed, substituting $W_{\mu\nu} = \Omega_{\mu\nu}^{(\mathrm{orb})} + \Omega_{\mu\nu}^{(\mathrm{F})}$ and using $P_{\mu\nu}U^\nu = 0$, one verifies that the Fermi part drops out under projection:

$$P_\mu^\alpha\, \Omega_{\alpha\beta}^{(\mathrm{F})}\, P_\nu^\beta = 0$$

because $\Omega_{\alpha\beta}^{(\mathrm{F})}$ contains at least one factor of U in each index position. Hence the projection isolates exactly the component of $W_{\mu\nu}$ that is orthogonal to U^μ, i.e. $\Omega_{\mu\nu}^{(\mathrm{orb})}$.

2.9.5 Fundamental decomposition of $W_{\mu\nu}$

According to above, the fundamental decomposition is

$$\begin{aligned} W_{\mu\nu} &= \Omega_{\mu\nu}^{(\mathrm{orb})} + \Omega_{\mu\nu}^{(\mathrm{F})} \\ \Omega_{\mu\nu}^{(\mathrm{orb})} &= W_{\mu\nu} - \Omega_{\mu\nu}^{(\mathrm{F})} \end{aligned}$$

which can be written succinctly as

$$W = \mathrm{P\,W\,P} + \frac{1}{\mathrm{c}^2}\,\mathrm{U} \wedge \alpha$$

$$W_{\mu\nu} = \underbrace{P_\mu^\alpha\, W_{\alpha\beta}\, P_\nu^\beta}_{\Omega_{\mu\nu}^{(\mathrm{orb})}} + \underbrace{\frac{1}{c^2}\,(U_\mu a_\nu - U_\nu a_\mu)}_{\Omega_{\mu\nu}^{(\mathrm{F})}}$$

The physical meaning is immediate:

- $\Omega_{\mu\nu}^{(\text{orb})}$ is the geometric rotation due to the worldline's orbital curvature,
- $\Omega_{\mu\nu}^{(\text{F})}$ collects the purely kinematical Fermi-Walker contribution.

2.9.6 Contraction with U^{μ}

Multiplying the decomposition by U^{ν} yields

$$\mathrm{W}_{\nu}^{\mu}U^{\nu} = \Omega_{\nu}^{(\text{orb})\mu}U^{\nu} - \Omega_{\nu}^{(\text{F})\mu}U^{\nu}$$

Since

$$\Omega^{(orb)}U = 0$$
$$\Omega^{(F)}U = a$$

the identity reduces to the universal dynamical law

$$a^{\mu} = \mathrm{W}_{\nu}^{\mu}\,U^{\nu}$$

Thus, the decomposition is perfectly consistent with the governing kinematic equation.

2.9.7 Three–dimensional form

Relation between frequencies

In the laboratory frame, write

$$U^{\mu} = (\gamma c, \gamma \mathbf{u}), \qquad a^{\mu} = \frac{dU^{\mu}}{d\tau}$$

The spatial part of the Fermi tensor is

$$\Omega_{ij}^{(\text{F})} = \frac{1}{c^2}\left(U_i a_j - U_j a_i\right) \Rightarrow \boldsymbol{\omega}_{\text{F}} = \frac{1}{2}\,\varepsilon_{ijk}\Omega_{jk}^{(\text{F})} = \frac{1}{c^2}\,(\mathbf{u} \times \mathbf{a})$$

Thus:

$$\boldsymbol{\omega}_{\mathrm{F}} = \frac{1}{c^2}(\mathbf{u} \times \mathbf{a})$$

This is the *Fermi (Thomas) precession vector* associated with the worldline.

As we know,

$$\boldsymbol{\xi} = \frac{1}{2}\varepsilon_{ijk}W_{jk} = \frac{1}{2}\varepsilon_{ijk}(\partial_j u_k - \partial_k u_j) = -\nabla \times \mathbf{u}$$

i.e. the kinematical vorticity vector of the congruence. Then

$$\boldsymbol{\xi} = \boldsymbol{\omega}_{\mathrm{orb}} + \boldsymbol{\omega}_{\mathrm{F}}$$

and since $\boldsymbol{\xi} = -\nabla \times \mathbf{u}$, we will have

$$\nabla \times \mathbf{u} = -\boldsymbol{\omega}_{\mathrm{orb}} - \boldsymbol{\omega}_{\mathrm{F}}$$

Astrovortex–free flow

In an astrovortex–free flow we have $\nabla \times \mathbf{u} = 0$ and therefore

$$\boldsymbol{\omega}_{\mathrm{orb}} = -\boldsymbol{\omega}_{\mathrm{F}} = -\frac{1}{c^2}(\mathbf{u} \times \mathbf{a})$$

This is the situation of a central–force orbital motion ($\mathbf{a} \parallel \mathbf{e}_r$, $\mathbf{u} = u_r\mathbf{e}_r + u_\phi\mathbf{e}_\phi$). Since

$$(\mathbf{u} \times \mathbf{a}) = u_\phi a_r\, \mathbf{e}_\theta \quad \Rightarrow \quad \boldsymbol{\omega}_{\mathrm{F}} = \frac{1}{c^2}(\mathbf{u} \times \mathbf{a})$$

the orbital piece becomes

$$\boldsymbol{\omega}_{\mathrm{orb}} = -\frac{1}{c^2}(\mathbf{u} \times \mathbf{a})$$

For central Newtonian gravity,

$$\mathbf{a} = -\frac{GM}{r^2}\,\mathbf{e}_r, \qquad \mathbf{u} = u_\phi\,\mathbf{e}_\phi, \qquad u_\phi = \frac{h}{r}$$

thus

$$\boldsymbol{\omega}_{\mathrm{F}} = \frac{1}{c^2}\left(\frac{h}{r}\mathbf{e}_\phi \times \left(-\frac{GM}{r^2}\mathbf{e}_r\right)\right) = +\frac{GMh}{c^2r^3}\mathbf{e}_z$$

Consequently:

$$\boldsymbol{\omega}_{\text{orb}} = -\frac{\mathrm{GMh}}{\mathrm{c}^2\mathrm{r}^3}\mathbf{e}_{\mathrm{z}}$$

As we will see in the next sections, the magnitude of $\boldsymbol{\omega}_{\text{orb}} \cdot \mathbf{h}$ coincide with the gravitomagnetic–type orbital potential term

$$\phi_{\text{orb}} \sim -\frac{GMh^2}{c^2r^3} = \boldsymbol{\omega}_{\text{orb}} \cdot \mathbf{h}$$

The time–space components of $\Omega_{\mu\nu}^{(orb)}$

The condition $\Omega^{(\text{orb})}U = 0$ implies, for the spatial sector,

$$\gamma\,\boldsymbol{\zeta} + \mathbf{u} \times \boldsymbol{\omega}_{\text{orb}} = 0$$

where $\boldsymbol{\zeta}$ encodes the mixed time–space components of $\Omega_{\mu\nu}^{(\text{orb})}$. Therefore

$$\boldsymbol{\zeta} = -\frac{1}{\gamma}\mathbf{u} \times \boldsymbol{\omega}_{\text{orb}}$$

and the full 4×4 structure of $\Omega_{\mu\nu}^{(\text{orb})}$ is completely determined by $\boldsymbol{\omega}_{\text{orb}}$ and $\mathbf{u}$.

2.9.8 Evolution of Geometric Objects under Fermi–Walker Transport

The Fermi–Walker tensor plays a distinguished geometric role in the relativistic evolution of vectors and tensors attached to a worldline. Let U^μ be the four–velocity of the material point, and let

$$\Omega_{\mu\nu}^{(\mathrm{F})} = \frac{1}{c^2}(a_\mu U_\nu - a_\nu U_\mu)$$

with $a^\mu = dU^\mu/d\tau$. This tensor satisfies the defining property

$$\Omega^{(F)}_{\mu\nu}\, U^{\nu} = a_{\mu}$$
$$\Omega^{(F)}_{\mu\nu}\, U^{\mu} = -a_{\nu}$$

and therefore, governs the "intrinsic" acceleration structure of the worldline.

Given any four–vector $A^{\mu}(\tau)$ which is *intrinsically fixed* in the rest frame of the particle (i.e. A'^{μ} = constant in the comoving frame), its evolution as seen in the laboratory frame obeys

$$\boldsymbol{\omega}_{\mathrm{F}} = \frac{1}{c^2}\,(\mathbf{u}\times\mathbf{a})$$

$$\frac{dA^{\mu}}{d\tau} = \Omega^{(F)\,\mu}{}_{\nu}\, A^{\nu}$$

Similarly, if $B^{\mu}{}_{\nu}(\tau)$ is a rank–2 tensor which is constant in the comoving rest frame (i.e. $B'^{\mu}{}_{\nu}$ = constant), then the laboratory observer detects the evolution

$$\frac{dB}{d\tau} = \Omega^{(F)}\, B - B\, \Omega^{(F)}$$

An important example is the orbital angular–momentum tensor $H_{\mu\nu}$. If $H'_{\mu\nu}$ is fixed in the local rest frame, then

$$\frac{dH}{d\tau} = \Omega^{(F)}\, H - H\, \Omega^{(F)}$$

In the 3–vector representation, the spatial part of Ω^{F} corresponds to the Thomas–like precession vector

$$\boldsymbol{\omega}_{\mathrm{F}} = \frac{1}{c^2}\,(\mathbf{u}\times\mathbf{a})$$

so that a purely spatial vector **A** fixed in the comoving frame evolves as

$$\frac{d\mathbf{A}}{d\tau} = \boldsymbol{\omega}_{\mathrm{F}}\times\mathbf{A}$$

Thus Fermi–Walker transport describes the unique non–rotating evolution along the worldline and acts as the geometric core of all intrinsic processional behavior in flat spacetime.

2.9.9 Closing Remark

The decomposition

$$W_{\mu\nu} = \Omega^{(\mathrm{orb})}_{\mu\nu} + \Omega^{(\mathrm{F})}_{\mu\nu}$$

$$\Omega^{(\mathrm{orb})} = PWP$$

where $P_{\mu\nu} = \eta_{\mu\nu} + U_\mu U_\nu / c^2$ is the spatial projector, reveals a clean separation between two conceptually distinct structures:

1. *Intrinsic Fermi–Walker geometry* ($\Omega^{(\mathrm{F})}$), which encodes the relativistic acceleration structure of the worldline and determines the evolution of all comoving vectors and tensors.

2. *Orbital kinematics* ($\Omega^{(\mathrm{orb})}$), obtained by projecting W into the spatial hyperplane orthogonal to U^μ. This isolates the purely spatial rotational content of W as perceived by the laboratory observer.

Together, $\Omega^{(\mathrm{F})}$ and $\Omega^{(\mathrm{orb})}$ reconstruct the full W, but each carries a different physical meaning: the former describes intrinsic, non–rotating evolution along the worldline, while the latter encodes the laboratory–measured orbital rotation of the trajectory. This clean geometric split will later play a fundamental role in the dynamical analysis of Flat Spacetime Gravity (FSG), where the interplay between intrinsic acceleration and projected orbital structure underlies both the kinetic equations and the emergence of effective potentials.

As will be clarified in the following subsection, this splitting is not specific to W but represents a direct application of a general geometric decomposition valid for any antisymmetric rank-2 tensor.

2.10 General Decomposition of Antisymmetric Tensors

2.10.1 Introductory Remark

The decomposition developed in the previous subsection for the kinematical tensor $W_{\mu\nu}$ is in fact a special instance of a general geometric identity. Any antisymmetric rank–2 tensor $A_{\mu\nu}$ can be uniquely decomposed into a purely spatial part (in the rest–frame of a given four–velocity V^{μ}) and a mixed time–space part determined by the contraction $A_{\mu\nu}V^{\nu}$. This subsection formulates the result in full generality.

2.10.2 Projector and notation

Let V^{μ} be any future–directed timelike unit four–velocity, $V_{\mu}V^{\mu} = -c^2$, and define the orthogonal projector

$$P(V)^{\mu}_{\nu} = \delta^{\mu}_{\nu} + \frac{1}{c^2}V^{\mu}V_{\nu}$$
$$P(V)^{\mu}_{\nu}V^{\nu} = 0$$

For any antisymmetric tensor $A_{\mu\nu} = -A_{\nu\mu}$ we denote by AV the four–vector

$$(AV)_{\mu} \equiv A_{\mu\nu}V^{\nu}$$

2.10.3 Statement of the decomposition

The tensor $A_{\mu\nu}$ admits the unique decomposition

$$A_{\mu\nu} = A^{(\mathrm{orb})}_{\mu\nu} + A^{(\mathrm{F})}_{\mu\nu}$$

Where

$$A^{(\mathrm{orb})}_{\mu\nu} = P(V)^{\alpha}_{\mu}\, A_{\alpha\beta}\, P(V)^{\beta}_{\nu}$$

$$A^{(\mathrm{F})}_{\mu\nu} = \frac{1}{c^2}\left(V_\mu (AV)_\nu - V_\nu (AV)_\mu\right) = \frac{1}{c^2} V \wedge (AV)$$

2.10.4 Properties

The two components satisfy the following identities:

$$A^{(\mathrm{orb})}_{\mu\nu} V^\nu = 0$$

$$A^{(\mathrm{F})}_{\mu\nu} V^\nu = (AV)_\mu$$

$$A^{(\mathrm{orb})}_{\mu\nu} = A_{\mu\nu} \quad \text{iff} \quad AV = 0$$

$$A^{(\mathrm{F})}_{\mu\nu} = 0 \quad \text{iff} \quad A_{\mu\nu} V^\nu = 0$$

Moreover, this decomposition of the antisymmetric tensor A is orthogonal with respect to the Minkowski inner product:

$$A^{(\mathrm{orb})}_{\mu\nu} A^{(\mathrm{F})\mu\nu} = 0$$

and is invariant under Lorentz transformations that leave V^μ fixed.

2.10.5 Specialization to the kinematical tensor $W_{\mu\nu}$

Setting

$$\begin{aligned} A_{\mu\nu} &= W_{\mu\nu} \\ V^\mu &= U^\mu \\ (AV)_\mu &= W_{\mu\nu} U^\nu = a_\mu \end{aligned}$$

the previous general formulas give

$$\begin{aligned} W^{(\mathrm{orb})}_{\mu\nu} &= P(U)^\alpha_\mu \, W_{\alpha\beta} \, P(V)^\beta_\nu \\ W^{(\mathrm{F})}_{\mu\nu} &= \frac{1}{c^2}\left(U_\mu a_\nu - U_\nu a_\mu\right) \end{aligned}$$

These coincide precisely with the tensors denoted Ω_{orb} and Ω_T in the previous subsection i.e.

$$W_{\mu\nu}^{(\mathrm{orb})} = \Omega_{\mu\nu}^{(\mathrm{orb})}$$

$$W_{\mu\nu}^{(\mathrm{F})} = \Omega_{\mu\nu}^{(\mathrm{F})}$$

2.10.6 Interpretation

The equity of decomposition extracts not only the purely spatial content of $A_{\mu\nu}$ as measured in the rest–frame of V^{μ}, but also the mixed time–space part that encodes how $A_{\mu\nu}$ acts on the four–velocity. Thus, the entire construction of the previous subsection follows as a direct special case of the general antisymmetric decomposition *of* this subsection.

2.11 Concluding Remarks

- The four–velocity generates a kinematic potential $\Phi_{\mu}^{(\mathrm{kin})} = -U_{\mu}$.
- Its derivatives form the antisymmetric tensor $W_{\mu\nu}$, with spatial content $(\boldsymbol{\delta}, \boldsymbol{\xi})$.
- The four–acceleration follows the exact identity $a_{\mu} = W_{\mu\nu} U^{\nu}$.
- The spatial acceleration obeys the relativistic hydrodynamic Euler equation.
- The angular–momentum bivector $H_{\mu\nu} = X \wedge U$ encodes the conserved specific angular momentum $\mathbf{h} = \mathbf{r} \times \mathbf{u}$ for central motion.
- The tensor $W_{\mu\nu}$ decomposes into orbital transport and Thomas–Fermi precession.
- Any antisymmetric tensor decomposes into orbital and Thomas–Fermi part.

This completes the purely kinematical framework. In the next sections, gravitational fields will be introduced in close analogy with electromagnetism, by coupling the matter current to a gravitational potential, while preserving the kinematic structure established above.

3. Flat-spacetime gravitational field

3.1 Introductory Remark

In this section we construct the linear flat-spacetime gravitational field theory of Flat Spacetime Gravity in direct analogy with classical electromagnetism, but strictly within flat Minkowski spacetime. Starting from a minimal electromagnetic to gravitational correspondence, we introduce a gravitational four–potential Φ_{μ}, the associated antisymmetric field tensor $G_{\mu\nu}$, and the spatial fields $\mathbf{g}$ (sourceful) and $\mathbf{b}$ (rotational), which play the roles of "gravitational electric" and "gravitational magnetic" fields respectively. The Maxwell–type equations for $(\mathbf{g}, \mathbf{b})$ are then derived, together with the Lorenz gauge condition and the flat–spacetime wave equations for Φ_{μ}. From the retarded potentials we obtain both static and quasi–static solutions, the Newtonian limit, and the leading velocity–dependent corrections. Gravitational waves, their energy–momentum content, and the exchange laws between field and matter are encoded in a symmetric stress–energy tensor that is quadratic in $G_{\mu\nu}$ and exactly parallels the electromagnetic case. Finally, the gravito-rotational sector is analyzed in detail through rotating and orbiting mass distributions, emphasizing how frame–dragging–like and spin–dependent astrophysical effects emerge here from a purely linear, Maxwell–type dynamics in a strictly Minkowskian background.

3.2 Electromagnetic to Gravitational Correspondence

We collect here the minimal dictionary of substitutions that allow the entire electromagnetic formalism to be transferred verbatim to flat–spacetime gravitation.

Constants

$$\varepsilon_0 \;\rightarrow\; \frac{1}{4\pi G}$$

$$\mu_0 \;\rightarrow\; \frac{4\pi G}{c^2}$$

Potentials and Field Tensor

$$A^{\mu}_{\text{EM}} \;\rightarrow\; \Phi^{\mu}$$
$$\phi_{\text{EM}} \;\rightarrow\; \phi$$
$$\mathbf{A}_{\text{EM}} \;\rightarrow\; \mathbf{A}$$

$$F_{\mu\nu} \;\rightarrow\; G_{\mu\nu}$$

Sources

$$q \;\rightarrow\; -m_0$$

$$\rho_{0,\text{EM}} \;\rightarrow\; -\rho_0$$
$$\rho_{\text{EM}} \;\rightarrow\; -\rho_0\,\gamma$$

$$J^{\mu}_{\text{EM}} \;\rightarrow\; -J^{\mu}$$
$$\mathbf{j}_{\text{EM}} \;\rightarrow\; -\mathbf{j}$$

Fields

$$\mathbf{E} \rightarrow \mathbf{g}$$
$$\mathbf{B} \rightarrow \mathbf{b}$$

Reminder Relations

$$J^{\mu}_{\text{EM}} = \rho_{0,\text{EM}}\,U^{\mu}$$
$$J^{\mu} = \rho_0\,U^{\mu}$$

$$\mathbf{j}_{\text{EM}} = \rho_{0,\text{EM}}\,\mathbf{u}$$
$$\mathbf{j} = \rho_0\,\mathbf{u}$$

Remark on sign conventions in the charge–mass correspondence

A subtle but crucial point in the electromagnetic–to–FSG dictionary concerns how the sign flip

$$q \;\rightarrow\; -m_0$$
$$\rho_{0,\text{EM}} \;\rightarrow\; -\rho_0$$

should be applied when electromagnetic expressions are translated into their gravitational analogues. The rule is this: *the sign change is applied once per physical interaction, not once per factor*. This reflects the fact that what physically corresponds between the two theories is not the algebraic symbol q but the interaction term as a whole.

For example, the Coulomb force

$$\frac{1}{4\pi\varepsilon_0}\frac{Q_1 Q_2}{r^2}$$

corresponds to the Newtonian gravitational force

$$-G\frac{M_1 M_2}{r^2}$$

with a *single* sign reversal. Likewise, the electromagnetic potential energy

$$\frac{1}{4\pi\varepsilon_0}\frac{Q_1 Q_2}{r}$$

maps to

$$-G\frac{M_1 M_2}{r}$$

Here the replacement

$$Q_1\, \phi_{(\mathrm{em})} \;\rightarrow\; M_1\, \phi_{(\mathrm{grav})}$$

already contains the required sign change, because the gravitational potential $\phi_{(\mathrm{grav})}$ is defined with the Newtonian minus sign. Thus, no additional minus sign should be applied to the source mass when forming the product. The sign convention is already encoded in the definition of the potential itself.

3.3 Gravitational potential and field tensor

3.3.1 Gravitational potential and field tensor

We begin by recalling the electromagnetic 4–potential,

$$A^{\mu} = \left(\frac{\Phi_{\mathrm{EM}}}{c}, \mathbf{A}_{\mathrm{EM}}\right)$$

and the electromagnetic field tensor

$$F_{\mu\nu} = \partial_{\mu}A_{\nu} - \partial_{\nu}A_{\mu}$$

Its components are

$$F_{0i} = -\frac{E_i}{c}$$
$$F_{ij} = +\varepsilon_{ijk} B_k$$

We introduce the gravitational 4–potential

$$\Phi^{\mu} = \left(\frac{\phi}{c}, \mathbf{A}\right)$$
$$\Phi_{\mu} = \left(-\frac{\phi}{c}, \mathbf{A}\right)$$

and define the gravitational field tensor

$$G_{\mu\nu} = \partial_{\mu}\Phi_{\nu} - \partial_{\nu}\Phi_{\mu}$$

The spatial fields are

$$\mathbf{g} = -\nabla\phi - \frac{\partial \mathbf{A}}{\partial t}$$
$$\mathbf{b} = \nabla \times \mathbf{A}$$

Thus

$$G_{0i} = -\frac{g_i}{c}$$
$$G_{ij} = +\varepsilon_{ijk} b_k$$

and the full 4 × 4 tensor is

$$G_{\mu\nu} = \begin{pmatrix} 0 & -\frac{g_x}{c} & -\frac{g_y}{c} & -\frac{g_z}{c} \\ +\frac{g_x}{c} & 0 & +b_z & -b_y \\ +\frac{g_y}{c} & -b_z & 0 & +b_x \\ +\frac{g_z}{c} & +b_y & -b_x & 0 \end{pmatrix}$$

The correspondence is structural:

$$(\Phi^{\mu}, G_{\mu\nu}) \leftrightarrow (A^{\mu}, F_{\mu\nu})$$
$$(\mathbf{g}, \mathbf{b}) \leftrightarrow (\mathbf{E}, \mathbf{B})$$

The vector field **g** may be interpreted as the *sourceful gravitational field*, while the vector field **b** represents the *rotational* gravitational field, in full analogy with the electric and magnetic fields of electromagnetism.

3.3.2 Invariants of the gravitational field tensor

A fundamental structural feature of any antisymmetric field tensor is the existence of two independent Lorentz–invariant scalars. For the gravitational tensor $G_{\mu\nu}$, defined by $G_{0i} = -g_i/c, \quad G_{ij} = +\varepsilon_{ijk} b_k$, the direct analogues of the electromagnetic invariants are

$$\mathrm{I}_1 \equiv \frac{1}{2} G_{\mu\nu} G^{\mu\nu} = |\mathbf{b}|^2 - \frac{|\mathbf{g}|^2}{c^2}$$

$$\mathrm{I}_2 \equiv \frac{1}{4} \varepsilon^{\mu\nu\alpha\beta} G_{\mu\nu} G_{\alpha\beta} = -\frac{1}{\mathrm{c}} \mathbf{g} \cdot \mathbf{b}$$

The invariant I_1 measures the "relative dominance" of the rotational field **b** over the sourceful field **g**; it changes sign when the configuration passes from sourceful–dominated ($|\mathbf{g}|/c > |\mathbf{b}|$) to rotationally–dominated ($|\mathbf{g}|/c < |\mathbf{b}|$) regimes. The pseudoscalar I_2 encodes the mutual alignment of **g** and **b**: it vanishes when the fields are orthogonal and takes its extremal values for parallel or antiparallel alignment. These two invariants fully classify the local algebraic structure of $G_{\mu\nu}$ under Lorentz transformations.

A further global invariant is the determinant of the full 4×4 antisymmetric matrix $G_{\mu\nu}$. For any such matrix the determinant takes the form

$$\det(G_{\mu\nu}) = \frac{1}{c^2}(\mathbf{g} \cdot \mathbf{b})^2 = I_2^2$$

which depends solely on the mixed invariant I_2. Thus $\det(G_{\mu\nu}) = 0$ whenever **g** and **b** are orthogonal, including all static and all plane–wave configurations; a nonzero determinant signals a nontrivial coupling between the sourceful and rotational parts of the gravitational field. These invariants therefore provide a compact and coordinate–independent characterization of the local structure of the FSG gravitational field, paralleling the well–known classification of electromagnetic fields.

3.4 Maxwell–Type Gravitational Field Equations

3.4.1 Homogeneous Maxwell–type equations

A central structural identity of any field tensor constructed as an exterior derivative of a four–potential Φ_μ is the Bianchi identity. Since the FSG gravitational tensor is defined by

$$G_{\mu\nu} = \partial_\mu \Phi_\nu - \partial_\nu \Phi_\mu$$

its exterior curvature automatically satisfies

$$\partial_\lambda G_{\mu\nu} + \partial_\mu G_{\nu\lambda} + \partial_\nu G_{\lambda\mu} = 0$$

This identity is purely geometric: it follows from the commutativity of partial derivatives and does *not* depend on the field equations or on the presence of sources. By projecting the index structure into temporal and spatial components, one immediately recovers the two homogeneous Maxwell–type equations of the Flat Spacetime Gravity theory. Taking $\lambda = 0$ and $\mu, \nu = i, j$ yields

$$\partial_0 G_{ij} + \partial_i G_{j0} + \partial_j G_{0i} = 0$$

and substituting $G_{0i} = -g_i/c$, $G_{ij} = \varepsilon_{ijk} b_k$ gives

$$\nabla \times \mathbf{g} = -\frac{\partial \mathbf{b}}{\partial t}$$

the gravitational analogue of Faraday's induction law. Similarly, choosing all indices spatial $(\lambda, \mu, \nu = i, j, k)$ yields

$$\partial_i G_{jk} + \partial_j G_{ki} + \partial_k G_{ij} = 0$$

which reduces directly to

$$\nabla \cdot \mathbf{b} = 0$$

which implies that the rotational field admits no monopole sources; in other words, no rotational gravitational monopoles exist.

Thus, exactly as in electromagnetism, the two homogeneous FSG equations do not constitute dynamical assumptions: they arise inevitably and uniquely from the Bianchi identity satisfied by the antisymmetric tensor $G_{\mu\nu}$, or equivalently from the statement that $G_{\mu\nu}$ is the exterior derivative of a four–potential Φ_μ.

3.4.2 Non-Homogeneous Maxwell–type equations

Using the correspondence rules, the electromagnetic Maxwell system in SI units,

$$\partial_\mu F^{\mu\nu} = \mu_0 J^\nu_{\mathrm{EM}}$$

maps to the gravitational system in the non-homogeneous Maxwell-type equation

$$\partial_\mu G^{\mu\nu} = -\frac{4\pi G}{c^2} J^\nu$$

Lorenz Gauge

Starting from the Maxwell–type field equation

$$\partial_\mu G^{\mu\nu} = -\frac{4\pi G}{c^2} J^\nu \qquad G_{\mu\nu} = \partial_\mu \Phi_\nu - \partial_\nu \Phi_\mu$$

we impose the gravitational Lorenz gauge

$$\partial_\mu \Phi^\mu \;=\; 0$$

The Lorenz condition couples the scalar and vector potentials:

$$\nabla \cdot \mathbf{A} + \frac{1}{c^2}\frac{\partial \phi}{\partial t} = 0$$

Under this gauge condition the identity

$$\partial_\mu \, \partial_\nu \Phi^\mu \;=\; \partial_\nu \, \partial_\mu \Phi^\mu \;= 0$$

by using the D' Alembert operator

$$\Box \;\equiv\; -\frac{1}{c^2}\frac{\partial^2}{\partial t^2} + \nabla^2$$

reduces to a d'Alembert equation for the potential:

$$\Box\, \Phi^\nu \;=\; -\frac{4\pi G}{c^2}\, J^\nu$$

Scalar and vector potential equation

Writing out the spatial and temporal components explicitly:

$$\Box\, \phi \;=\; -\,4\pi G\, \rho_0 \gamma$$

$$\Box\, \mathbf{A} \;=\; -\frac{4\pi G}{c^2}\, \mathbf{j}$$

The structure is formally identical to electromagnetism:

- The source of the scalar potential is the *relativistic density* $\rho_0\gamma$

- The source of the vector potential is the *mass current* $\mathbf{j} = \rho_0 \mathbf{u}$

- The propagation operator is the flat-spacetime d'Alembertian: $\Box = -c^{-2}\,\partial_t^2 + \nabla^2$

Thus, the gravitational potentials ϕ and $\mathbf{A}$ propagate as waves at the speed of light *in strictly flat Minkowski spacetime*, exactly mirroring the electromagnetic case.

Non-homogeneous Maxwell–type equations in 3-vector form

As we have seen, the sourceful and the rotational gravitational fields are:

$$\mathbf{g} = -\nabla\phi - \frac{\partial \mathbf{A}}{\partial t}$$
$$\mathbf{b} = \nabla \times \mathbf{A}$$

Substituting these fields into the covariant non-homogeneous Maxwell–type equation yields the 3-vector form:

$$\nabla \cdot \mathbf{g} = -4\pi G\, \rho_0 \gamma$$
$$\nabla \times \mathbf{b} - \frac{1}{c^2}\frac{\partial \mathbf{g}}{\partial t} = -\frac{4\pi G}{c^2}\mathbf{j}$$

Because the proper mass density ρ_0 plays the role of the proper charge density in electromagnetism, the corresponding laboratory-frame density is $\rho_0\gamma$, and therefore the gravitational Gauss law carries the factor γ exactly as shown above.

3.4.3 All 3-vector Maxwell–type equations

$$\nabla \cdot \mathbf{b} = 0$$

$$\nabla \times \mathbf{g} = -\frac{\partial \mathbf{b}}{\partial t}$$

$$\nabla \cdot \mathbf{g} = -4\pi G\, \rho_0 \gamma$$

$$\nabla \times \mathbf{b} - \frac{1}{c^2}\frac{\partial \mathbf{g}}{\partial t} = -\frac{4\pi G}{c^2}\,\mathbf{j}$$

3.4.4 Integral form of the Maxwell–type equations

Absence of gravito-rotational monopoles

The identity $\nabla \cdot \mathbf{b} = 0$ yields the flux relation

$$\oint_S \mathbf{b} \cdot \hat{\mathbf{n}}\, dS = 0$$

for any closed surface and, as we have seen, excluding the possibility of isolated rotational–monopole sources in the Flat Spacetime Gravity framework.

Integral Gauss-type Law

From the equation

$$\nabla \cdot \mathbf{g} = -4\pi G\, \rho_0 \gamma$$

we get that for any closed surface S with outward unit normal $\hat{\mathbf{n}}$,

$$\oint_S \mathbf{g} \cdot \hat{\mathbf{n}}\, dS = -4\pi G \int_V \rho_0\, \gamma\, dV$$

where V is the volume enclosed by S. The flux of the sourceful gravitational field $\mathbf{g}$ through a closed surface is therefore proportional to the total relativistic mass contained inside, establishing the exact analogue of Gauss' law for the electric field.

Thus, for the exterior of a symmetric body with mass M we will have:

$$\mathbf{g}(\mathrm{r}) = -\frac{\mathrm{G\ M}}{\mathrm{r}^2}\,\hat{\mathbf{r}}$$

Integral Ampère-type Law

For any open surface S with boundary curve ∂S,

$$\oint_{\partial S} \mathbf{b} \cdot d\boldsymbol{\ell} = -\frac{4\pi G}{c^2}\int_S \mathbf{j} \cdot \hat{\mathbf{n}}\, dS + \frac{1}{c^2}\frac{d}{dt}\int_S \mathbf{g} \cdot \hat{\mathbf{n}}\, dS$$

The line integral of the rotational field **b** around a closed loop is thus determined by two contributions: the flux of mass current **j** (the analogue of electric current) through the surface, and the time variation of the **g**-flux (the gravitational analogue of the displacement current).

These integral laws show that the gravitational field in the Flat Spacetime Gravity obeys precisely the same flux and circulation structures as electromagnetism, with mass density and mass current replacing charge and electric current, and with the sourceful field **g** and rotational field **b** assuming the roles of "gravitational electric" and "gravitational magnetic" components.

3.4.5 A Hidden $G_{\mu\nu} - W_{\mu\nu}$ Coupling in Flat Spacetime Gravity

The $G_{\mu\nu} - W_{\mu\nu}$ connection

A remarkable structural identity emerges when we combine the inhomogeneous Maxwell–type equation

$$\partial_\mu G^{\mu\nu} = -\frac{4\pi G}{c^2} J^\nu$$

with the definition of the mass–current four–vector $J^\nu = \rho_0 U^\nu$ for a compact body with constant rest density ρ_0. Taking one more derivative we obtain

$$\partial^\mu G_{\mu\nu} = -\frac{4\pi G}{c^2} J_\nu$$

$$\partial_\mu J_\nu = \rho_0\, \partial_\mu U_\nu$$

Applying the d'Alembert operator to $G_{\mu\nu}$ and using the inhomogeneous equation yields the flat–spacetime wave equation

$$\Box\, G_{\mu\nu} = \frac{4\pi G}{c^2}\, \rho_0\, W_{\mu\nu}$$

where

$$W_{\mu\nu} \equiv \partial_\nu U_\mu - \partial_\mu U_\nu$$

is precisely the vorticity tensor of the four–velocity field.

Proof

$$\partial_\mu G^\mu_\nu = -\frac{4\pi G}{c^2} J_\nu$$

$$\partial_\alpha G_{\mu\nu} + \partial_\mu G_{\nu\alpha} + \partial_\nu G_{\alpha\mu} = 0$$

$$\Box\, X_{\mu\nu} \equiv \eta^{\alpha\beta}\, \partial_\alpha\, \partial_\beta X_{\mu\nu}$$

$$\eta^{\alpha\beta}\, \partial_\beta \left(\partial_\alpha G_{\mu\nu} + \partial_\mu G_{\nu\alpha} + \partial_\nu G_{\alpha\mu}\right) = 0$$

$$\eta^{\alpha\beta}\, \partial_\beta\, \partial_\alpha G_{\mu\nu} + \eta^{\alpha\beta}\, \partial_\beta\, \partial_\mu G_{\nu\alpha} + \eta^{\alpha\beta}\, \partial_\beta\, \partial_\nu G_{\alpha\mu} = 0$$

$$\Box\, G_{\mu\nu} + \partial_\mu\left(\eta^{\alpha\beta}\, \partial_\beta G_{\nu\alpha}\right) + \partial_\nu\left(\eta^{\alpha\beta}\, \partial_\beta G_{\alpha\mu}\right) = 0$$

$$\partial_\gamma G^\gamma_\nu = -\frac{4\pi G}{c^2} J_\nu$$

$$\eta^{\alpha\beta}\, \partial_\beta G_{\nu\alpha} = \partial_\beta G^\beta_\nu = +\frac{4\pi G}{c^2} J_\nu$$

$$\eta^{\alpha\beta}\, \partial_\beta G_{\alpha\mu} = -\eta^{\alpha\beta}\, \partial_\beta G_{\mu\alpha} = -\,\partial_\beta G^\beta_\mu = -\frac{4\pi G}{c^2} J_\mu$$

$$\Box\, G_{\mu\nu} + \partial_\mu\left(+\frac{4\pi G}{c^2} J_\nu\right) + \partial_\nu\left(-\frac{4\pi G}{c^2} J_\mu\right)$$

$$\Box\, G_{\mu\nu} + \frac{4\pi G}{c^2}\, \partial_\mu J_\nu - \frac{4\pi G}{c^2}\, \partial_\nu J_\mu = 0$$

$$\Box\, G_{\mu\nu} = -\frac{4\pi G}{c^2}(\partial_\mu J_\nu - \partial_\nu J_\mu)$$

$$\Box\, G_{\mu\nu} = -\frac{4\pi G}{c^2}\rho_0(\partial_\mu U_\nu - \partial_\nu U_\mu)$$

$$\Box\, G_{\mu\nu} = +\frac{4\pi G}{c^2}\rho_0(\partial_\nu U_\mu - \partial_\mu U_\nu)$$

$$\Box\, G_{\mu\nu} = +\frac{4\pi G}{c^2}\rho_0\, W_{\mu\nu}$$

Alternative Derivation via Commutation of the d'Alembertian and the 4D Curl

In Minkowski spacetime the partial derivatives commute, so the d'Alembert operator

$$\Box = \partial_\alpha\, \partial^\alpha$$

commutes with any first–order differential operator built from partial derivatives. In particular, for any four–vector field A_μ we have

$$\begin{aligned} \Box(\mathrm{Grad}(A)) &= \mathrm{Grad}(\Box A) \\ \Box(\mathrm{Div}(A)) &= \mathrm{Div}(\Box A) \\ \Box(\mathrm{Curl}(A)) &= \mathrm{Curl}(\Box\, A) \end{aligned}$$

where “Grad”, “Div” and “Curl” denote the appropriate 4D differential operators.

By construction the FSG field tensor $G_{\mu\nu}$ is the 4D curl of the potential Φ_μ,

$$G_{\mu\nu} = (\mathrm{Curl}\,\Phi)_{\mu\nu} = \partial_\mu\Phi_\nu - \partial_\nu\Phi_\mu$$

while the kinematical tensor $W_{\mu\nu}$ is (up to a sign) the 4D curl of the four–velocity,

$$W_{\mu\nu} = -\,(\mathrm{Curl}\,U)_{\mu\nu} = -(\partial_\mu U_\nu - \partial_\nu U_\mu)$$

Assuming, for a compact and homogeneous source, that the rest–mass density ρ_0 is constant in the region under consideration, the potential obeys the inhomogeneous wave equation

$$\Box\,\Phi_\mu = -\frac{4\pi G}{c^2}\,\rho_0\,U_\mu$$

Applying the d'Alembertian to $G_{\mu\nu}$ and using commutation with the 4D curl, we obtain

$$\Box\,G_{\mu\nu} = \Box(\mathrm{Curl}\,\Phi)_{\mu\nu} = (\mathrm{Curl}\,(\Box\Phi))_{\mu\nu} = \mathrm{Curl}\left(-\frac{4\pi G}{c^2}\,\rho_0\,U\right)_{\mu\nu}$$

With ρ_0 constant this reduces to

$$\Box\,G_{\mu\nu} = -\frac{4\pi G}{c^2}\,\rho_0\,(\mathrm{Curl}\,U)_{\mu\nu} = +\frac{4\pi G}{c^2}\,\rho_0\,W_{\mu\nu}$$

where in the last step we used $W_{\mu\nu} = -(\mathrm{Curl}\,U)_{\mu\nu}$. Thus we recover, in a single line,

$$\Box\,G_{\mu\nu} = \frac{4\pi G}{c^2}\,\rho_0\,W_{\mu\nu}$$

in full agreement with the more detailed derivation based on the Bianchi identity and the inhomogeneous Maxwell equation.

Hidden connection in 3D form

If we apply the relation

$$\Box\,G_{\mu\nu} = \frac{4\pi G}{c^2}\,\rho_0\,W_{\mu\nu}$$

to its components, we obtain

$$\Box\mathbf{g} = \frac{4\pi G}{c^2}\rho_0\,\boldsymbol{\delta}$$

$$\Box\mathbf{b} = \frac{4\pi G}{c^2}\rho_0\,\boldsymbol{\xi}$$

Given that

$$\boldsymbol{\delta} = c^2\,\nabla\gamma + \frac{\partial\mathbf{u}}{\partial t}, \qquad \boldsymbol{\xi} = -\nabla\times\mathbf{u}$$

we will finally have:

$$\Box\mathbf{g} = 4\pi G\rho_0\,\nabla\gamma$$

$$\square \mathbf{b} \;=\; -\frac{4\pi G}{c^2}\rho_0 \,\nabla\times\mathbf{u} \;=\; -\frac{4\pi G}{c^2}\,\nabla\times\mathbf{j}$$

Summary

This identity reveals a structural coupling that is absent or highly obscured in the curved–metric formulation of general relativity: the *sourceful flat–spacetime gravitational field* $G_{\mu\nu}$ is driven not only by mass density, but also directly by the vorticity of the four–velocity field. Thus, rotational flow of matter (encoded in $W_{\mu\nu}$) acts as a covariant source for wave excitations of $G_{\mu\nu}$. In the Flat Spacetime Gravity formulation this connection appears naturally and linearly, exposing a clean Minkowskian symmetry between the two antisymmetric tensors

$$G_{\mu\nu} \quad \leftrightarrow \quad W_{\mu\nu}$$

3.5 Retarded potentials

The causal (retarded) solution of $\square\,\Phi^\nu = -4\pi G\, J^\nu/c^2$ is

$$\Phi^\nu(t,\mathbf{x}) = -\frac{G}{c^2}\int \frac{J^\nu(t_{\text{ret}},\mathbf{x}')}{|\mathbf{x}-\mathbf{x}'|}\,d^3x', \qquad t_{\text{ret}} \,=\, t - \frac{|\mathbf{x}-\mathbf{x}'|}{c}$$

Writing explicitly:

$$\phi(t,\mathbf{x}) \;=\; -G\int \frac{\rho_0(t_{\text{ret}},\mathbf{x}')}{|\mathbf{x}-\mathbf{x}'|}\,d^3x'$$

$$\mathbf{A}(t,\mathbf{x}) \;=\; -\frac{G}{c^2}\int \frac{\mathbf{j}(t_{\text{ret}},\mathbf{x}')}{|\mathbf{x}-\mathbf{x}'|}\,d^3x'$$

The gravitational fields follow from

$$\mathbf{g} = -\nabla\phi - \frac{\partial\mathbf{A}}{\partial t}$$

$$\mathbf{b} = \nabla\times\mathbf{A}$$

so that

$$\mathbf{g}(t,\mathbf{x}) = +G\int \rho_0(t_{\text{ret}},\mathbf{x}')\,\nabla_{\mathbf{x}}\left(\frac{1}{|\mathbf{x}-\mathbf{x}'|}\right) d^3x' + \frac{\partial}{\partial t}\left[\frac{G}{c^2}\int \frac{\mathbf{j}(t_{\text{ret}},\mathbf{x}')}{|\mathbf{x}-\mathbf{x}'|} d^3x'\right]$$

$$\mathbf{b}(t,\mathbf{x}) = -\nabla\times\left[\frac{G}{c^2}\int \frac{\mathbf{j}(t_{\text{ret}},\mathbf{x}')}{|\mathbf{x}-\mathbf{x}'|} d^3x'\right]$$

In the static limit $\mathbf{u} = 0, \mathbf{j} = 0$, one recovers

$$\phi(\mathbf{x}) = -G\int \frac{\rho_0(\mathbf{x}')}{|\mathbf{x}-\mathbf{x}'|} d^3x'$$

$$\mathbf{A} = \mathbf{0}$$

so,

$$\mathbf{b} = \mathbf{0}$$

3.6 Static Fields for a Moving Source

3.6.1 Scalar and vector potentials and induced sourceful and rotational fields

Consider a point mass M whose rest–density is $\rho_0(\mathbf{x}) = M\,\delta^3(\mathbf{x})$ and which moves with constant velocity $\mathbf{u}_s$. The retarded potentials reduce, in the quasi–static approximation where the motion is slow on the light–crossing time of the system, to a remarkably simple structure.

To leading order in u_s/c,

$$\phi(t,\mathbf{x})\& = -G\,\frac{M}{R}\left[1+\mathcal{O}\left(\frac{u_s^2}{c^2}\right)\right], \qquad R = |\mathbf{x}-\mathbf{x}_s(t)|$$

$$\mathbf{A}(t,\mathbf{x})\& = -\frac{G}{c^2}\,\frac{M\,\mathbf{u}_s}{R}\left[1+\mathcal{O}\left(\frac{u_s^2}{c^2}\right)\right]$$

Thus, even a mass in uniform rectilinear motion generates a nonzero rotational–type potential $\mathbf{A}$ and hence a nonzero gravito-rotational field

$\mathbf{b} = \nabla \times \mathbf{A}$. This is the precise analogue of the electromagnetic vector potential of a uniformly moving charge.

Using $\mathbf{g} = -\nabla\phi - \partial\mathbf{A}/\partial t$ and $\mathbf{b} = \nabla \times \mathbf{A}$, one obtains, for motion at constant $\mathbf{u}_s$,

$$\mathbf{g}(t,\mathbf{x}) = -GM\frac{\mathbf{R}}{R^3}\left[1 + \mathcal{O}\left(\frac{u_s^2}{c^2}\right)\right]$$
$$\mathbf{b}(t,\mathbf{x}) = \frac{GM}{c^2}\frac{\mathbf{u}_s \times \mathbf{R}}{R^3}\left[1 + \mathcal{O}\left(\frac{u_s^2}{c^2}\right)\right]$$

Hence:

- the sourceful field $\mathbf{g}$ retains the Newtonian $1/R^2$ structure at leading order,
- while the rotational field $\mathbf{b}$ is suppressed by a factor u_s/c.

3.6.2 Static Gravitational Fields

For a strictly static matter distribution, both the density ρ_0 and the spatial current $\mathbf{j}$ are time–independent, and the retarded potentials reduce to their instantaneous form. Specifically,

$$\phi(\mathbf{x}) = -G\int \frac{\rho_0(\mathbf{x}')}{|\mathbf{x} - \mathbf{x}'|}\, d^3x'$$
$$\mathbf{A}(\mathbf{x}) = \mathbf{0}$$

since a static configuration has $\mathbf{j} = \mathbf{0}$ everywhere. Consequently,

$$\mathbf{b}(\mathbf{x}) = \nabla \times \mathbf{A}(\mathbf{x}) = \mathbf{0}$$

and the only non–vanishing gravitational field is the sourceful field

$$\mathbf{g}(\mathbf{x}) = -\nabla\phi(\mathbf{x}) = -G\int \rho_0(\mathbf{x}')\,\nabla\left(\frac{1}{|\mathbf{x} - \mathbf{x}'|}\right) d^3x'$$

This expression is valid for *any* static mass distribution, and it provides the universal Newtonian form of the gravitational field in the present flat–spacetime framework. In particular, for a point mass M located at the origin one obtains immediately

$$\phi(r) = -\frac{GM}{r}$$

$$\mathbf{g}(r) = -\frac{GM}{r^2}\,\hat{\mathbf{r}}$$

Because the Maxwell–type system (3.3) is linear in the sources ρ_0 and $\mathbf{j}$, static gravitational fields obey an exact superposition principle:

$$\phi = \phi_1 + \phi_2 + \cdots$$

$$\mathbf{g} = \mathbf{g}_1 + \mathbf{g}_2 + \cdots.$$

This contrasts sharply with general relativity, where nonlinear couplings prevent superposition except in special cases. Here, linearity is preserved because spacetime remains strictly Minkowskian and the gravitational field does not source itself.

3.6.3 Static Spherical Symmetry

We now specialize the Maxwell–type equations of Flat Spacetime Gravity to a static, spherically symmetric source. Assume:

$$\mathbf{j} = 0, \qquad \partial_t \rho_0 = 0$$

$$\phi = \phi(r), \qquad \mathbf{A} = 0$$

Since $\mathbf{A} = 0$,

$$\mathbf{b} = \nabla \times \mathbf{A} = 0$$

Thus, a static spherical source produces no gravito-rotational field.

The static Maxwell–type equation reduces to the Poisson equation

$$\nabla^2 \phi = -4\pi G\,\rho_0$$

Outside the source, Gauss' law gives the unique spherically symmetric solution

$$\phi(r) = -\frac{GM}{r}$$

Using $\mathbf{g} = -\nabla\phi$,

$$\mathbf{g}(r) = -\frac{d}{dr}\left(-\frac{GM}{r}\right)\hat{\mathbf{r}} = -\frac{GM}{r^2}\hat{\mathbf{r}}$$

Thus, for any static, spherically symmetric source:

$$\phi(r) = -\frac{GM}{r}, \qquad \mathbf{A} = 0$$

$$\mathbf{g}(r) = -\frac{GM}{r^2}\hat{\mathbf{r}}, \qquad \mathbf{b} = 0$$

This is the unique static spherical solution of the flat–spacetime Maxwell–type system of Flat Spacetime Gravity.

3.7 Gravitational Waves in the linear FSG Framework

3.7.1 Wave Equations in Vacuum

In vacuum, the flat-spacetime gravitational field inherits the full wave structure of electromagnetism through the replacement

$$A^{\mu} \rightarrow \Phi^{\mu}, \qquad F_{\mu\nu} \rightarrow G_{\mu\nu}, \qquad \Box A^{\mu} = 0 \rightarrow \Box \Phi^{\mu} = 0$$

where

$$\Box \equiv -\frac{1}{c^2}\frac{\partial^2}{\partial t^2} + \nabla^2$$

In this section we collect the corresponding wave equations, plane-wave solutions, orthogonality conditions, energy–flux relations, polarizations, and the wave interpretation in terms of the potential itself.

Using the relations

$$\Box \mathbf{g} = 4\pi G \rho_0 \nabla\gamma$$

$$\square\,\mathbf{b} \;=\; -\frac{4\pi G}{c^2}\rho_0\,\nabla\times\mathbf{u} \;=\; -\frac{4\pi G}{c^2}\,\nabla\times\mathbf{j}$$

we obtain for the vacuum ($\rho_0 = 0$)

$$\square\mathbf{g} = 0, \qquad \square\mathbf{b} = 0$$

$$\nabla^2\mathbf{g} - \frac{1}{c^2}\frac{\partial^2\mathbf{g}}{\partial t^2} = 0$$

$$\nabla^2\mathbf{b} - \frac{1}{c^2}\frac{\partial^2\mathbf{b}}{\partial t^2} = 0$$

Thus both **g** and **b** satisfy standard wave equations in flat Minkowski spacetime, propagating at the speed c.

3.7.2 Plane-Wave Solutions

A plane-wave solution of the potential equation in vacuum can be written as

$$\Phi^\mu(x) = \Re\{\varepsilon^\mu\, e^{i(k_\nu x^\nu)}\}$$

$$k_\nu x^\nu = -\omega t + \mathbf{k}\cdot\mathbf{x}$$

where $k^\mu = (\omega/c, \mathbf{k})$ is the wave four-vector and ε^μ is a constant polarization four-vector. The wave equation $\square\,\Phi^\mu = 0$ implies the null condition

$$k_\mu k^\mu = -\frac{\omega^2}{c^2} + |\mathbf{k}|^2 = 0 \quad \Rightarrow \quad \omega = c\,|\mathbf{k}|$$

The Lorenz gauge condition $\partial_\mu\Phi^\mu = 0$ yields the transversality constraint

$$k_\mu\varepsilon^\mu = 0$$

In a radiation gauge (for example $\phi = 0$ and $\nabla\cdot\mathbf{A} = 0$) one may take

$$\Phi^\mu(x) = \Re\{(0, \mathbf{A}_0)\, e^{i(\mathbf{k}\cdot\mathbf{x}-\omega t)}\}$$

$$\mathbf{k}\cdot\mathbf{A}_0 = 0$$

Then the corresponding fields are

$$\mathbf{g}(x) = -\frac{\partial \mathbf{A}}{\partial t} = \Re\{\, i\omega\, \mathbf{A}_0\, e^{i(\mathbf{k}\cdot\mathbf{x}-\omega t)}\}$$

$$\mathbf{b}(x) = \nabla \times \mathbf{A} = \Re\{\, i\,\mathbf{k} \times \mathbf{A}_0\, e^{i(\mathbf{k}\cdot\mathbf{x}-\omega t)}\}$$

3.7.3 Orthogonality Relations

Let

$$\mathbf{g}(x) = \Re\{\mathbf{g}_0\, e^{i(\mathbf{k}\cdot\mathbf{x}-\omega t)}\}$$
$$\mathbf{b}(x) = \Re\{\mathbf{b}_0\, e^{i(\mathbf{k}\cdot\mathbf{x}-\omega t)}\}$$

with $\omega = c|\mathbf{k}|$. From the Maxwell–type equations in vacuum we obtain

$$\mathbf{k} \cdot \mathbf{g}_0 = 0$$
$$\mathbf{k} \cdot \mathbf{b}_0 = 0$$

and

$$\mathbf{k} \times \mathbf{g}_0 = \omega\, \mathbf{b}_0$$
$$\mathbf{k} \times \mathbf{b}_0 = -\frac{\omega}{c^2}\mathbf{g}_0$$

These relations imply that $\mathbf{g}_0$, $\mathbf{b}_0$, and $\mathbf{k}$ are mutually orthogonal, forming a right-handed triad. Moreover, taking magnitudes, we find

$$|\mathbf{b}_0| = \frac{|\mathbf{k} \times \mathbf{g}_0|}{\omega} = \frac{|\mathbf{k}|}{\omega}\,|\mathbf{g}_0| = \frac{1}{c}\,|\mathbf{g}_0|$$

so that

$$|\mathbf{g}_0| = c\,|\mathbf{b}_0|$$

Thus, gravitational waves in this linear FSG formulation are transverse waves with $\mathbf{g}\perp\mathbf{b}\perp\mathbf{k}$, fully analogous to the transverse electromagnetic waves.

3.7.4 Energy Density, Poynting Vector, and Continuity

The energy density and Poynting vector are defined as

$$u_{\text{field}} = \frac{c^2}{8\pi G}\left(\frac{|\mathbf{g}|^2}{c^2} + |\mathbf{b}|^2\right)$$

$$\mathbf{S} = \frac{c^2}{4\pi G}\,\mathbf{g} \times \mathbf{b}$$

The continuity equation follows directly from Maxwell–type equations:

$$\frac{\partial u_{\text{field}}}{\partial t} + \nabla \cdot \mathbf{S} = -\mathbf{g} \cdot \mathbf{j}$$

In vacuum:

$$\frac{\partial u_{\text{field}}}{\partial t} + \nabla \cdot \mathbf{S} = 0$$

For a monochromatic plane wave, the energy density is proportional to $|\mathbf{g}_0|^2/c^2 + |\mathbf{b}_0|^2$, and the Poynting vector is directed along the wave vector **k**, describing energy transport at the speed c.

3.7.5 Polarizations

A plane wave propagating along, say, the z-axis,

$$\mathbf{k} = (0,0,k), \qquad \omega = ck$$

admits two independent transverse polarization states. Choosing an orthonormal basis $(\hat{\mathbf{e}}_x, \hat{\mathbf{e}}_y, \hat{\mathbf{e}}_z)$, a general plane wave can be written as

$$\mathbf{g}(t,\mathbf{x}) = \Re\{(g_x\,\hat{\mathbf{e}}_x + g_y\,\hat{\mathbf{e}}_y)e^{i(kz-\omega t)}\}$$

$$\mathbf{b}(t,\mathbf{x}) = \Re\{(b_x\,\hat{\mathbf{e}}_x + b_y\,\hat{\mathbf{e}}_y)e^{i(kz-\omega t)}\}$$

with the constraint $|\mathbf{g}_0| = c|\mathbf{b}_0|$ and $\mathbf{b}_0 = \hat{\mathbf{k}} \times \mathbf{g}_0/c$. Linear polarizations correspond to $\mathbf{g}_0$ aligned along $\hat{\mathbf{e}}_x$ or $\hat{\mathbf{e}}_y$; circular and elliptical polarizations are obtained by taking complex linear combinations of these two basis states, exactly as in electromagnetism.

3.7.6 Potential Waves and Gauge Structure

Since $\square\, \Phi^{\mu} = 0$ in vacuum, the potential itself carries a plane-wave structure:

$$\Phi^{\mu}(x) = \Re\{\varepsilon^{\mu}\, e^{i(k_{\nu}x^{\nu})}\}, \qquad k_{\mu}k^{\mu} = 0, \qquad k_{\mu}\varepsilon^{\mu} = 0$$

Two potentials related by a gauge transformation

$$\Phi^{\mu} \;\rightarrow\; \Phi^{\mu} + \partial^{\mu}\Lambda$$

with $\square\, \Lambda = 0$ represent the same physical **g** and **b**. In radiation gauge ($\phi = 0$, $\nabla \cdot \mathbf{A} = 0$) the physical degrees of freedom reduce precisely to the two transverse polarizations described above. The wave nature of Φ^{μ} is therefore completely consistent with the transverse, massless propagation of the flat-spacetime gravitational field at speed c, within the strictly flat Minkowski background of Flat Spacetime Gravity.

3.7.7 Flat Spacetime Gravity Produces Waves Easily

In Flat Spacetime Gravity the wave character is automatic: No curved metric, no perturbations, no weak–field limits. The linear, Minkowskian structure yields gravitational waves exactly like EM. The emergence of gravitational waves is not a subtle or higher–order phenomenon but an immediate structural consequence of the theory itself.

Because the field tensor $G_{\mu\nu}$ is defined exactly as in electromagnetism and obeys the same homogeneous Bianchi identity together with the same Lorenz–type gauge condition, the flat–spacetime d'Alembert equation for the potential, $\square\, \Phi_{\mu} = 0$, appears automatically and without approximation whenever the sources vanish. No perturbative expansion of a background metric is required, no weak–field linearisation is invoked, and no curvature terms need to be retained or discarded.

The very act of replacing $(A_{\mu}, F_{\mu\nu})$ by $(\Phi_{\mu}, G_{\mu\nu})$ in Minkowski space reproduces at once the entire radiative sector: transverse waves, two independent polarizations, energy flux via a Poynting–type vector, and a symmetric field stress–energy tensor. Thus, the existence, structure, and propagation of gravitational radiation are not additional hypotheses of the theory but a direct inheritance of the Maxwell–type formulation; the wave character of the gravitational field is revealed clearly by recognizing that the sourceful and rotational components $(\mathbf{g}, \mathbf{b})$ obey the same geometric equations as their electromagnetic counterparts, even in a strictly flat spacetime.

3.8 Energy–Momentum Tensor

3.8.1 Field Energy–Momentum Tensor

By strict analogy with electromagnetism, the gravitational field tensor

$$G_{\mu\nu} = \partial_\mu \Phi_\nu - \partial_\nu \Phi_\mu$$

admits the quadratic, symmetric energy–momentum tensor

$$T^{\mu\nu}_{\text{field}} = \frac{c^2}{4\pi G}\left(G^{\mu}_{\ \rho}\, G^{\nu\rho} - \frac{1}{4}\eta^{\mu\nu} G_{\alpha\beta} G^{\alpha\beta}\right)$$

where $\eta^{\mu\nu} = \mathrm{diag}(-1,1,1,1)$. The overall factor $c^2/4\pi G$ plays the role of the electromagnetic $1/\mu_0$.

Using

$$G_{0i} = -\frac{g_i}{c}, \qquad G_{ij} = +\varepsilon_{ijk}\, b_k$$

one finds the energy density and the Poynting vector

$$u_{\text{field}} = T^{00}_{\text{field}} = \frac{c^2}{8\pi G}\left(\frac{|\mathbf{g}|^2}{c^2} + |\mathbf{b}|^2\right)$$

$$T^{0i}_{\text{field}} = \frac{c}{4\pi G}(\mathbf{g} \times \mathbf{b})_i = \frac{1}{c} S_i$$

$$\mathbf{S} = c\, T^{0i}_{\text{field}} = \frac{c^2}{4\pi G}\, \mathbf{g} \times \mathbf{b}$$

as we saw in the waves section and in exact analogy with electromagnetism.

3.8.2 Matter Energy–Momentum Tensor

For a dust source with rest–mass density ρ_0 and four–velocity U^μ,

$$T^{\mu\nu}_{\text{matter}} = \rho_0 \, U^\mu U^\nu$$

The conserved gravitational four–current is

$$J^\mu = \rho_0 U^\mu = (\rho_0 \gamma c, \ \rho_0 \mathbf{u})$$

so that the spatial current density is simply

$$\mathbf{j} = \rho_0 \mathbf{u}$$

3.8.3 Exchange Law Between Field and Matter

The inhomogeneous Maxwell–type equations read

$$\partial_\mu G^{\mu\nu} = \frac{4\pi G}{c^2} J^\nu$$

and by direct differentiation we obtain the exact interaction identity

$$\partial_\mu T^{\mu\nu}_{\text{field}} = -\frac{1}{c^2} \, G^\nu_\rho \, J^\rho$$

$$\partial_\mu T^{\mu\nu}_{\text{matter}} = +\frac{1}{c^2} \, G^\nu_\rho \, J^\rho$$

Therefore,

$$\partial_\mu \left(T^{\mu\nu}_{\text{field}} + T^{\mu\nu}_{\text{matter}} \right) = 0$$

which is the exact local conservation of total energy–momentum in the flat-spacetime gravitational theory.

3.8.4 Energy Continuity and Work Law

Taking $\nu = 0$ yields the energy equation of the Poynting vector:

$$\frac{\partial u_{\text{field}}}{\partial t} + \nabla \cdot \mathbf{S} = -\mathbf{g} \cdot \mathbf{j}$$

showing that the power transferred to matter is

$$\mathcal{P}_{\text{matter}} = \mathbf{g} \cdot \mathbf{j}$$

In vacuum ($\mathbf{j} = 0$):

$$\frac{\partial u_{\text{field}}}{\partial t} + \nabla \cdot \mathbf{S} = 0$$

3.8.5 Momentum Continuity

The spatial components $\nu = i$ give the momentum–balance law:

$$\frac{\partial}{\partial t}\left(\frac{\mathbf{S}}{c^2}\right) + \nabla \cdot \boldsymbol{\sigma} = -\rho_0 \mathbf{g}$$

where the stress tensor is

$$\sigma_{ij} = T^{ij}_{\text{field}} = \frac{c^2}{4\pi G}\left(-\frac{g_i g_j}{c^2} - b_i b_j + \frac{1}{2}\delta_{ij}\left[\frac{|\mathbf{g}|^2}{c^2} + |\mathbf{b}|^2\right]\right)$$

This structure is the exact analogue of the electromagnetic Maxwell stress tensor.

3.8.6 Pressure and Internal Stresses

In Newtonian and relativistic continuum dynamics the spatial components of the stress–energy tensor encode not only momentum fluxes but also internal mechanical stresses. In the Flat Spacetime Gravity framework the field stress tensor T^{ij}_{field} represents the "gravitational" part of these stresses. When matter possesses internal isotropic pressure p (e.g. a fluid or stellar interior), the total stress–energy tensor must incorporate this contribution in the usual flat–spacetime form

$$T^{\mu\nu}_{\text{matter}} = (\rho_0 c^2 + \varepsilon + p)\, U^\mu U^\nu + p\, \eta^{\mu\nu}$$

where ε is the internal energy density. The spatial components reduce to

$$T^{ij}_{\text{matter}} = (\rho_0 c^2 + \varepsilon + p)\, U^i U^j + p\, \delta^{ij}$$

showing that the isotropic pressure appears as the isotropic diagonal part of the stress tensor, while anisotropic stresses (shear) arise from nonzero products $U^i U^j$.

The total stress–energy tensor is the sum of matter and field contributions,

$$T^{\mu\nu}_{\text{tot}} = T^{\mu\nu}_{\text{matter}} + T^{\mu\nu}_{\text{field}}$$

and the Flat Spacetime Gravity conservation law

$$\partial_\mu T^{\mu\nu}_{\text{tot}} = 0$$

leads simultaneously to conservation of energy ($\nu = 0$) and momentum ($\nu = i$), with pressure gradients acting as explicit force-density terms:

$$\partial_t\left[(\rho_0 c^2 + \varepsilon + p)\gamma^2 v^i\right] + \partial_j\left[T^{ij}_{\text{matter}} + T^{ij}_{\text{field}}\right] = 0$$

Thus, in the Flat Spacetime Gravity formulation pressure plays exactly the same dynamical role as in relativistic hydrodynamics: it is a source of momentum flux and a contributor to the stress balance, fully compatible with the flat-spacetime conservation equations and the Maxwell-type structure of the theory.

3.8.7 Summary

The field energy–momentum tensor of Flat Spacetime Gravity is:

completely symmetric, divergenceless when combined with matter, quadratic *in* $G_{\mu\nu}$, structurally identical to electromagnetism, and naturally supports energy flux, radiation pressure, and gravitational-wave transport.

3.9 Gravito-rotational Field of Rotating Mass Distributions

3.9.1 Dipole Approximation for a Rotating Body

As we have seen, the gravito-rotational vector potential generated by a mass current distribution,

$$\mathbf{A}(\mathbf{x}) = -\frac{G}{c^2}\int \frac{\mathbf{j}(\mathbf{x}')}{|\mathbf{x}-\mathbf{x}'|}\, d^3x'$$

The associated gravito-rotational field is

$$\mathbf{b} = \nabla \times \mathbf{A}$$

For a compact rotating source whose size is small compared with the observation radius $r = |\mathbf{x}|$, one defines the mass-current dipole moment

$$\mathbf{J} = \int \mathbf{x}' \times \mathbf{j}(\mathbf{x}')\, d^3x'$$

Using the standard electromagnetic to gravitational replacement rule

$$\frac{\mu_0}{4\pi}\,\boldsymbol{\mu}_{\mathrm{EM}} \;\to\; \frac{G}{c^2}\,\mathbf{J}$$

the far–zone potential becomes

$$\mathbf{A}(\mathbf{x}) = -\frac{G}{c^2}\frac{\mathbf{J}\times\mathbf{r}}{r^3}, \qquad \mathbf{r} = \mathbf{x}$$

The corresponding field is

$$\mathbf{b}(\mathbf{x}) \;=\; -\frac{G}{c^2}\left[\frac{3(\mathbf{J}\cdot\hat{\mathbf{r}})\hat{\mathbf{r}} - \mathbf{J}}{r^3}\right]$$

3.9.2 Axisymmetric Rigidly Rotating Spheres

For a rigid sphere of mass M, radius R and uniform density rotating with angular velocity $\boldsymbol{\Omega}$,

$$\mathbf{J} = \frac{2}{5} M R^2 \mathbf{\Omega}$$

and for all $r > R$,

$$\mathbf{b}(\mathbf{x}) = -\frac{G}{c^2}\left[\frac{3(\mathbf{J}\cdot\hat{\mathbf{r}})\hat{\mathbf{r}} - \mathbf{J}}{r^3}\right]$$

3.9.3 Circular Mass Currents

Consider a thin ring of radius a carrying constant mass current I_m. On the symmetry axis:

$$A_\varphi(0,0,z) = -\frac{G}{c^2}\frac{I_m a^2}{(a^2 + z^2)^{3/2}}$$

For $z \gg a$,

$$A_\varphi \approx -\frac{G}{c^2}\frac{J}{z^2}$$

$$J = I_m \pi a^2$$

3.9.4 Orbiting Point Mass (Nonrelativistic Limit)

For a point mass in circular orbit of radius a and speed $v \ll c$,

$$\mathbf{J} = mav\,\hat{\mathbf{z}}$$

and the far–zone field reduces to

$$\mathbf{b} = -\frac{G}{c^2}\left[\frac{3(\mathbf{J}\cdot\hat{\mathbf{r}})\hat{\mathbf{r}} - \mathbf{J}}{r^3}\right]$$

3.9.5 Remarks on Astrophysical Consequences

The gravito-rotational dipole field of Flat Spacetime Gravity provides a unified flat–spacetime language for several observational effects that are usually discussed within curved–metric formalisms.

- First, frame–dragging–type phenomena arise naturally from the coupling of test–particle angular momentum to the dipole field $\mathbf{b}(\mathbf{x}) = -\frac{G}{c^2}[3(\mathbf{J}\cdot\hat{\mathbf{r}})\hat{\mathbf{r}} - \mathbf{J}]/r^3$, leading to secular precessions of orbital planes and gyroscope axes that mimic the Lense–Thirring effect but are here encoded purely in the Maxwell–type rotational sector.
- Second, gravitational lensing receives subleading spin–dependent corrections: the deflection angle acquires a small contribution proportional to the projection of $\mathbf{J}$ onto the impact–parameter direction, so that co–rotating and counter–rotating light rays experience slightly different bending, again without invoking spacetime curvature.
- Third, in compact binaries with at least one rapidly rotating component, the gravito-rotational field modulates both orbital phases and signal propagation times; in pulsar–timing applications this appears as characteristic, spin–dependent timing residuals that can in principle be computed directly from the flat–spacetime FSG equations.

Because the FSG rotational field follows a Maxwell-type structure, these phenomena emerge from well understood flat-space dynamics, independent of curved-spacetime formalisms. Thus, a single linear, Minkowskian framework accounts simultaneously for frame dragging, lensing corrections, and timing effects, with all spin–induced signatures traced back to the same dipolar mass–current structure.

3.10 Gravito-rotational Larmor Frequency

In the Flat Spacetime Gravity framework, a rotating test body placed in an external gravito-rotational field **b** and undergoing accelerated motion with four–velocity U^μ and four–acceleration α^μ experiences a combined precession of its intrinsic angular momentum. The purely kinematic part of this precession is encoded in the Thomas–type bivector

$$\Omega^{(T)}_{\mu\nu} = \frac{1}{c^2}(\alpha^\mu U^\nu - \alpha^\nu U^\mu)$$

which is the opposite of the Fermi-Walker tensor $\Omega^{(F)}_{\mu\nu}$

$$\Omega^{(T)}_{\mu\nu} = -\,\Omega^{(F)}_{\mu\nu}$$

In 3D form, its spatial component is

$$\boldsymbol{\omega}_{\mathrm{Th}} = \frac{1}{c^2}\,\boldsymbol{\alpha}\times\mathbf{u} = -\frac{1}{c^2}\,\mathbf{u}\times\mathbf{a} = -\boldsymbol{\omega}_F$$

with $\mathbf{u} = \gamma\mathbf{v}$ the spatial part of U^μ. This is the Thomas–type precession vector: even in flat Minkowski spacetime, a spinning body with nonzero acceleration $\boldsymbol{\alpha}$ and velocity $\mathbf{u}$ experiences a purely kinematic precession at the rate $\boldsymbol{\omega}_{\mathrm{Th}}$.

In addition, and in complete analogy with electromagnetism, a rotating test body placed in an external gravito-rotational field $\mathbf{b}$ experiences a torque that drives a steady precession of its intrinsic angular momentum. Let $\mathbf{S}$ denote the spin (or internal angular momentum) of the test body, and $\mathbf{b}$ the external FSG rotational field generated by distant sources. The coupling follows directly from the dipole–field interaction term, we have already seen:

$$\boldsymbol{\tau} = \frac{G}{c^2}\,\mathbf{S}\times\mathbf{b}$$

which is the exact analogue of the electromagnetic torque $\boldsymbol{\tau}_{\mathrm{EM}} = \boldsymbol{\mu}\times\mathbf{B}$ for a magnetic moment $\boldsymbol{\mu}$. This torque induces a precession of the spin vector with angular velocity

$$\boldsymbol{\omega}_{\mathrm{L}} = -\,\mathbf{b}$$

the *gravitational Larmor frequency*. Thus, a gyroscope, a spinning particle, or any rigid body with intrinsic angular momentum will precess uniformly around the direction of $\mathbf{b}$ at a rate proportional to the local gravito-rotational field, even in perfectly flat spacetime.

Thus, the total precession rate becomes

$$\boldsymbol{\omega}_{\rm tot} = \boldsymbol{\omega}_{\rm L} + \boldsymbol{\omega}_{\rm F} = \boldsymbol{\omega}_{\rm L} - \boldsymbol{\omega}_{\rm T} = -\,\mathbf{b} + \frac{1}{c^2}\,\mathbf{u} \times \mathbf{a}$$

Exactly as in electromagnetism, the first term represents dynamical coupling to the external rotational field, while the second arises purely from the acceleration of the moving frame. For bound orbits, this leads to a secular spin–precession effect that parallels the de Sitter/Thomas precession of general relativity, but here emerges entirely from the Maxwell–type structure of Flat Spacetime Gravity without curvature or perturbative expansions.

Consequently, rotating test bodies act as sensitive probes of the gravito-rotational sector: their steady precession encodes both the local magnitude and direction of **b**, while motion–induced effects introduce corrections proportional to $\mathbf{u} \times \mathbf{a}$. The gravitational Larmor frequency therefore provides a direct operational signature of gravito- rotational fields in the Flat Spacetime Gravity framework, completing the analogy with the electromagnetic dynamics of spinning dipoles and offering a clear physical interpretation of the **b**–field as a generator of rotational inertial effects in flat spacetime.

3.11 Concluding Remarks

The analysis presented in this section demonstrates that the gravito–rotational structure of Flat Spacetime Gravity arises entirely from the Maxwell–type formulation on flat Minkowski spacetime, without invoking curved metrics, weak–field expansions, or perturbative assumptions. Once the correspondence

$$(A_\mu, F_{\mu\nu}) \;\rightarrow\; (\Phi_\mu, G_{\mu\nu})$$

is established, the entire kinematical dynamics follows directly from the antisymmetry of $G_{\mu\nu}$ and the Bianchi identity. The entire gravitational dynamics follows from the inhomogeneous field equations, describing how mass density and mass current generate the sourceful and rotational gravitational fields $(\mathbf{g}, \mathbf{b})$, while the homogeneous equations encode their geometric constraints. These structures are not imposed; they are inevitable consequences of the tensorial definition itself.

Within this unified framework, the static Newtonian limit emerges directly from the $\nabla \cdot \mathbf{g}$ equation, whereas the gravito-rotational sector arises from mass currents and rotational motion in exact analogy with electromagnetism. The existence of gravitational waves is likewise automatic: in the absence of sources, the potentials satisfy the flat–spacetime wave equation, leading to transverse propagating modes, a Poynting–type flux vector, and a symmetric stress–energy tensor that mirrors the electromagnetic one. Thus, the radiative, dynamical, and energetic content of the theory is already encoded in the linear, Minkowskian structure, without any appeal to spacetime curvature.

Spin dynamics further reinforces the natural completeness of the framework. A rotating test body in an external **b**–field undergoes Larmor–type precession with angular frequency $-\mathbf{b}$, while accelerated motion induces a Thomas–type kinematic precession.

Taken together, these results reveal a coherent and economic picture: all gravitational phenomena investigated here—static fields, mass–current effects, radiation, precession, energy and momentum transport—arise from a single linear field tensor on flat spacetime. What in curved–metric formalisms are attributed to geometric deformation (frame dragging, gravitational waves, spin precession, anisotropic stresses) emerges in Flat Spacetime Gravity solely from the properties of $(\Phi_\mu, G_{\mu\nu})$ and the dynamics of mass currents.

In this sense, the gravito–rotational field of Flat Spacetime Gravity is not a geometric curvature but a genuine Maxwell–type physical field living in Minkowski space. The theory achieves a unified treatment of gravitational phenomena through the simplest possible tensorial structure, providing a transparent and predictive platform on which the subsequent sections will build the full orbital, astrophysical, and observational consequences of the gravito–rotational interaction.

4. Effective Potential

4.1 Introductory Remark

In the previous section we constructed the fundamental flat-spacetime gravitational field $G_{\mu\nu}$ generated by a source through Maxwell–type relations. What remains is to understand how a test body, following a worldline and carrying its own rotational state, *responds* to that field.

This section introduces a new and essential idea: the *orbital potential* $\Phi^{(\mathrm{orb})\mu}$. This object provides the missing coupling between the external field of the source and the motion of the test particle. It depends simultaneously on $G_{\mu\nu}$ and on the kinematical bivector $\widehat{H}^{\mu\nu}$ of the particle, and therefore represents a genuinely new ingredient — a dynamical interaction term that ties the geometry of the field to the geometry of the trajectory itself.

Adding this orbital contribution to the fundamental potential yields the *effective* potential and, from it, the effective field tensor. These in turn lead to a Lorentz–type dynamical law, a modified radial acceleration, and a full set of conservation theorems that include the Newtonian limit and its relativistic corrections.

This section develops these ideas step by step: the construction of $\Phi^{(\mathrm{orb})\mu}$, the effective potential and field, the resulting equation of motion, and the associated conserved quantities. Together they establish how the flat-spacetime gravitational field acts on matter and how orbital dynamics arise naturally within flat spacetime.

4.2 The Projected Rotational Tensor $\widehat{H}^{\mu\nu}$

4.2.1 The angular momentum tensor

The kinematical angular–momentum tensor (per unit mass) of a test body is defined by

$$\mathrm{H}^{\mu\nu} = \mathrm{X}^{\mu}\,\mathrm{U}^{\nu} - \mathrm{X}^{\nu}\,\mathrm{U}^{\mu}$$

with X^{μ} the position four–vector and U^{μ} the four–velocity. Its spatial block reproduces the three–angular–momentum vector

$$\mathbf{h} = \mathbf{r} \times \mathbf{u}$$

but $H^{\mu\nu}$ also contains time–space components that have no direct operational meaning for the source.

4.2.2 The need for projection

A central physical point is that the gravitational source, with four–velocity N^{μ}, can only "measure" spatial angular momentum in its own rest frame. In Minkowski spacetime, the tangent space at each event splits uniquely into

$$T = \mathrm{Span}(N^{\mu}) \oplus \Sigma, \qquad \Sigma = \{V^{\mu}: N_{\mu}V^{\mu} = 0\}$$

the one–dimensional time axis of the source and its orthogonal three–dimensional spatial hyperplane. Only vectors and bivectors lying in Σ correspond to physically observable spatial quantities for that source. Therefore, using the full $H^{\mu\nu}$ — which mixes the time direction of the source with orbital degrees of freedom of the probe — would produce frame–dependent, non–covariant coupling terms.

To restore Lorentz–covariant spatial coupling, one must project $H^{\mu\nu}$ onto Σ and having defined

$$\mathrm{H}^{\mu\nu} = \mathrm{X}^{\mu}\mathrm{U}^{\nu} - \mathrm{X}^{\nu}\mathrm{U}^{\mu}$$

we now apply the standard orthogonal projector

$$P(V)_{\mu\nu} = \eta_{\mu\nu} + \frac{V_{\mu}V_{\nu}}{c^2}$$

onto the rest–space of the source, so $V^{\mu} = N^{\mu}$:

$$P^{\mu\nu} = \eta^{\mu\nu} + \frac{N^{\mu}N^{\nu}}{c^2}$$
$$P^{\mu\nu}N_{\nu} = 0$$
$$P^{\mu}_{\alpha}\, P^{\alpha\nu} = P^{\mu\nu}$$

4.2.3 Definition of the projected rotational tensor.

We therefore introduce

$$\widehat{H}^{\mu\nu} = P^{\mu}_{\alpha} H^{\alpha\beta} P^{\nu}_{\beta}$$

This tensor is:

- antisymmetric,
- Lorentz–covariant,
- spatial with respect to the source:

$$\widehat{H}^{\mu\nu} N_{\nu} = 0$$

- and reduces in the rest frame of the source to a pure 3×3 angular–momentum tensor representing **h**.

In this sense, $\widehat{H}^{\mu\nu}$ is not an arbitrary modification of $H^{\mu\nu}$: it is the *unique Lorentz–covariant spatial part of* $H^{\mu\nu}$ *relevant for the interaction with the source*. Any alternative would be frame–dependent and would not define a physically meaningful coupling.

The tensor $\widehat{H}^{\mu\nu} = P^{\mu}_{\alpha} H^{\alpha\beta} P^{\nu}_{\beta}$ represents the projection of the angular–momentum tensor $H^{\mu\nu}$ onto the three–dimensional rest–space orthogonal to the four–velocity N^{μ}. This projected tensor encodes precisely the part of $H^{\mu\nu}$ that is purely spatial for an observer comoving with the source. In the *instantaneous rest frame* of the source, where $N^{\mu} = (c, 0,0,0)$ and consequently $P_{\mu\nu} = \mathrm{diag}(0,1,1,1)$, the 3×3 spatial block of $\widehat{H}^{\mu\nu}$ coincides exactly with the spatial block of $H^{\mu\nu}$. In a general laboratory frame, however, the spatial components differ, causing $\widehat{H}_{,ij} \neq H_{ij}$ unless evaluated in the instantaneous rest frame of the source. Thus, $\widehat{H}^{\mu\nu}$ should be interpreted as the *source's rest–frame spatial content* of $H^{\mu\nu}$, not merely its spatial block in an arbitrary coordinate system.

4.2.4 Explicit 4×4 form in the source rest frame

In the instantaneous rest frame of the source ($N^\mu = (c, 0,0,0)$), the projector becomes

$$P^\mu_{\ \nu} = \begin{pmatrix} 0 & 0 & 0 & 0 \\ 0 & 1 & 0 & 0 \\ 0 & 0 & 1 & 0 \\ 0 & 0 & 0 & 1 \end{pmatrix}$$

and therefore

$$\widehat{H}^{\mu\nu} = \begin{pmatrix} 0 & 0 & 0 & 0 \\ 0 & 0 & +h_z & -h_y \\ 0 & -h_z & 0 & +h_x \\ 0 & +h_y & -h_x & 0 \end{pmatrix}$$

Thus $\widehat{H}^{0\mu} = 0$ and the entire physical content of the tensor is encoded in the spatial pseudovector $\mathbf{h} = (h_x, h_y, h_z)$. This is precisely the information to which the source can couple.

The tensor $\widehat{H}^{\mu\nu}$ will serve as the core ingredient for constructing the orbital gravitational potential in the next subsection.

4.3 The Orbital Potential $\Phi^{(\mathrm{orb})}$

4.3.1 General form of the orbital potential

The orbital gravitational potential introduces a covariant coupling between the external field $G_{\mu\nu}$ of the source and the kinematical structure of the test body, encoded in $\widehat{H}^{\mu\nu}$ and U^μ. It acts along the worldline of the particle and constitutes a genuine "orbital'' potential: it depends explicitly on the motion of the probe and its angular momentum, and therefore *cannot* be generated by Maxwell–type field equations alone, which only relate $G_{\mu\nu}$ to external sources.

Any gravitational four–potential decomposes as

$$\Phi^\mu = \left(\frac{\phi}{c}, \mathbf{A}\right)$$

hence, we define

$$\Phi^{(\mathrm{orb})\mu} = \left(\frac{\phi^{(\mathrm{orb})}}{c},\ \mathbf{A}^{(\mathrm{orb})}\right)$$

4.3.2 Covariant definition

The unique Lorentz–covariant orbital potential is

$$\Phi^{(\mathrm{orb})\mu} = \frac{1}{c^2}\, G^{\mu\rho}\, \widehat{H}_{\rho\nu}\, U^{\nu}$$

This is linear in $G_{\mu\nu}$, bilinear in $(\widehat{H}, U)$, and transforms as a four–vector. It realises the desired coupling between the source field and the kinematical bivector of the test body along its worldline.

4.3.3 Temporal component

Since $\Phi^{(\mathrm{orb})0} = \phi^{(\mathrm{orb})}/c$, we obtain

$$\frac{\phi^{(\mathrm{orb})}}{c} = \frac{1}{c^2}\, G^{0\rho}\, \widehat{H}_{\rho\nu}\, U^{\nu}$$

and therefore

$$\phi^{(\mathrm{orb})} = \frac{1}{c}\, G^{0\rho}\, \widehat{H}_{\rho\nu}\, U^{\nu}$$

In the rest frame of the source, $G^{0i} = +g^i/c$ and only the spatial block of $\widehat{H}_{ij}$ contributes. Using $\widehat{H}_{ij} = \varepsilon_{ijk}\widehat{h}_k$,

$$\phi^{(\mathrm{orb})} = \frac{1}{c}\frac{g^i}{c}\left(\varepsilon_{ijk} h_k\right) u^j = \frac{1}{c^2}\, \mathbf{g}\cdot(\mathbf{u}\times\mathbf{h})$$

For a central field, $\mathbf{g} = -GM\,\hat{\mathbf{r}}/r^2$ and $\mathbf{h} = \mathbf{r}\times\mathbf{u}$, $\mathbf{u}_\varphi = \mathrm{h}/\mathrm{r}$ leading to

$$\phi^{(\mathrm{orb})} = -\frac{GM\,h^2}{c^2 r^3}$$

Geometric Interpretation: $\phi^{(orb)}$ as an Oriented Spatial Volume

A noteworthy structural property of the orbital potential is that its defining contraction

$$\phi^{(\mathrm{orb})} = \frac{1}{c^2}\,\mathbf{g}\cdot(\mathbf{u}\times\mathbf{h})$$

is a mixed triple product of three spatial vectors. In three–dimensional Euclidean geometry one has the identity

$$\mathbf{g}\cdot(\mathbf{u}\times\mathbf{h}) = \det(\mathbf{g},\mathbf{u},\mathbf{h})$$

so that

$$\phi^{(\mathrm{orb})} = \frac{1}{c^2}\det(\mathbf{g},\mathbf{u},\mathbf{h})$$

Hence $\phi^{(\mathrm{orb})}$ is not a conventional potential derived from a scalar field; instead, it represents the *oriented spatial volume* of the parallelepiped spanned by the gravitational acceleration **g**, the velocity **u**, and the orbital angular–momentum vector **h**. Its sign reverses under any inversion of the orientation of these vectors, and its magnitude is proportional to the geometric volume they enclose.

Using

$$\mathbf{g} = -\frac{GM}{r^2}\mathbf{e}_r, \qquad \mathbf{h} = \mathbf{r}\times\mathbf{u}$$

the triple product reduces to

$$\det(\mathbf{g},\mathbf{u},\mathbf{h}) = -\frac{GM}{r^2}\,\mathbf{e}_r\cdot\big(\mathbf{u}\times(\mathbf{r}\times\mathbf{u})\big) = -\frac{GMh^2}{r^3}$$

yielding

$$\phi^{(\mathrm{orb})} = -\frac{GMh^2}{c^2r^3}$$

in perfect agreement with the expression obtained earlier.

This geometric viewpoint reveals that the orbital potential is a genuinely three–dimensional, orientation–sensitive dynamical correction. It has no Newtonian analogue because it arises not from a scalar gravitational field but from the spatial configuration of motion inside that field. In this sense, $\phi^{(\mathrm{orb})}$ encodes a geometric coupling between the gravitational acceleration, the particle's velocity, and its orbital angular momentum—an intrinsically flat–spacetime, first–order relativistic effect.

4.3.4 Spatial components and geometric vanishing

The spatial part is

$$A^{(\mathrm{orb})i} = \frac{1}{c^2}\, G^{i\rho}\, \widehat{H}_{\rho\nu}\, U^{\nu}$$

Geometrically, $\widehat{H}_{\rho\nu}$ is confined to the spatial hyperplane orthogonal to N^{μ}:

$$\widehat{H}_{0\nu} = 0, \qquad \widehat{H}_{\mu\nu} N^{\mu} = 0$$

On the other hand, $G^{\mu\nu}$ is an antisymmetric two–form and $G^{i\rho}$ selects a mixed time–space component when contracted with U_{ν}. The contraction with $\widehat{H}_{\rho\nu}$ then attempts to contract a mixed (time–space) wedge with a purely spatial bivector. Because all time–space components of $\widehat{H}_{\rho\nu}$ vanish, this mixed contraction identically vanishes:

$$A^{(orb)i} \;=\; 0$$

4.3.5 Final form

Thus

$$\Phi^{(\mathrm{orb})\mu} \;=\; \left(\frac{\phi^{(\mathrm{orb})}}{c},\, \mathbf{0}\right), \qquad \phi^{(\mathrm{orb})} \;=\; -\,\frac{GM\,h^2}{c^2 r^3}$$

The potential $\Phi_{\mu}^{(\mathrm{orb})}$ therefore provides a covariant and physically motivated orbital correction to the fundamental potential Φ_{μ}, encoding the coupling between the source field $G_{\mu\nu}$ and the orbital angular momentum of the test body along its worldline. It will be added to Φ_μ in the next subsection to form the effective gravitational potential $\Phi_\mu^{(\text{eff})}$.

4.4 The Effective Potential $\Phi^{(\text{eff})}$

4.4.1 Component form

The physical gravitational interaction experienced by a test body in motion is determined not only by the fundamental potential Φ^μ generated by the source, but also by the orbital term $\Phi^{(\text{orb})\mu}$ which encodes the coupling between the external field $G_{\mu\nu}$ and the kinematical structure $(\widehat{H}^{\mu\nu}, U^\mu)$ of the particle along its worldline. The natural object governing the dynamics is therefore the *effective* four–potential

$$\Phi^{(\text{eff})\mu} \quad = \quad \Phi^\mu + \Phi^{(\text{orb})\mu}$$

Using the decompositions

$$\Phi^\mu = \left(\frac{\phi}{c}, \mathbf{A}\right)$$

$$\Phi^{(\text{orb})\mu} \quad = \quad \left(\frac{\phi^{(\text{orb})}}{c}, \mathbf{A}^{(\text{orb})}\right)$$

we obtain the explicit structure

$$\Phi^{(\text{eff})\mu} = \left(\frac{\phi + \phi^{(\text{orb})}}{c}, \mathbf{A} + \mathbf{A}^{(\text{orb})}\right)$$

Thus, the orbital correction contributes solely to the scalar part of the potential, leaving the spatial part unchanged:

$$\phi^{(\text{eff})} \;=\; \phi + \phi^{(\text{orb})}$$

$$\mathbf{A}^{(\text{eff})} \;=\; \mathbf{A} + \mathbf{A}^{(\text{orb})}$$

Since $\mathbf{A}^{(\text{orb})} = 0$ we will finally have

$$\phi^{(\text{eff})} \;=\; \phi + \phi^{(\text{orb})}$$

$$\mathbf{A}^{(\mathrm{eff})} = \mathbf{A}$$

4.4.2 Explicit orbital correction

From the previous subsection, the orbital scalar potential is

$$\phi^{(\mathrm{orb})} = -\frac{GM\,h^2}{c^2 r^3}$$

where $\mathbf{h} = \mathbf{r} \times \mathbf{u}$ is the instantaneous orbital angular–momentum vector of the test body, thus

$$\phi^{(eff)} = \phi - \frac{GM\,h^2}{c^2 r^3}$$

4.4.3 Final expression

The effective potential takes the compact and physically transparent form

$$\Phi^{(\mathrm{eff})\mu} = \left(\frac{\phi - \frac{GM\,h^2}{c^2 r^3}}{c}, \mathbf{A} \right)$$

This potential incorporates both the Newtonian–like attraction encoded in ϕ and the kinematical orbital coupling encoded in $\phi^{(\mathrm{orb})}$. It is this combined potential that will generate, in the next subsection, the effective field tensor $G_{\mu\nu}^{(\mathrm{eff})}$ and the corresponding equation of motion.

4.5 The Effective Field Tensor $G^{(\mathrm{eff})}$

4.5.1 Construction of the Effective Field Tensor

Once the effective potential

$$\Phi^{(\mathrm{eff})\mu} = \left(\frac{\phi^{(\mathrm{eff})}}{c}, \mathbf{A} \right)$$

$$\phi^{(\mathrm{eff})} = \phi + \phi^{(\mathrm{orb})}$$

is defined, the corresponding field tensor is obtained in complete analogy with gravitational tensor $G_{\mu\nu}$ through the standard linear antisymmetric construction

$$G_{\mu\nu}^{(\mathrm{eff})} = \partial_\mu \Phi_\nu^{(\mathrm{eff})} - \partial_\nu \Phi_\mu^{(\mathrm{eff})}$$

Note that, because the definition is linear in Φ^μ, we have immediately

$$G_{\mu\nu}^{(\mathrm{eff})} = G_{\mu\nu} + G_{\mu\nu}^{(\mathrm{orb})}$$

with the orbital contribution

$$G_{\mu\nu}^{(\mathrm{orb})} = \partial_\mu \Phi_\nu^{(\mathrm{orb})} - \partial_\nu \Phi_\mu^{(\mathrm{orb})}$$

4.5.2 Time–space block

Since $\Phi^{(\mathrm{orb})\mu}$ contains only a temporal part,

$$\Phi^{(\mathrm{orb})\mu} = \left(\frac{\phi^{(\mathrm{orb})}}{c}, \mathbf{0}\right)$$

The sourceful gravitational field, however, acquires an additional term,

$$g_i^{(\mathrm{eff})} = g_i + g_i^{(\mathrm{orb})}$$

with

$$g_i^{(\mathrm{orb})} = -\,\partial_i \phi^{(\mathrm{orb})} = -\,\partial_i\left(-\frac{GM\,h^2}{c^2 r^3}\right) = +\,\partial_i\left(\frac{GM\,h^2}{c^2 r^3}\right)$$

So

$$g_i^{(\mathrm{eff})} = g_i + \partial_i\left(\frac{GM\,h^2}{c^2 r^3}\right)$$

4.5.3 Spatial block

Since $\mathbf{A}^{(\mathrm{orb})} = \mathbf{0}$ identically, the spatial part of the effective vector potential is unchanged and the gravito-rotational field $\mathbf{b}$ receives no orbital correction:

$$G_{ij}^{(\mathrm{orb})} = 0, \qquad b_k^{(\mathrm{eff})} = b_k$$

$$\mathbf{b}^{(\mathrm{eff})} = \mathbf{b}$$

Thus, the only modification of the field tensor induced by the orbital coupling is contained in the time–space block.

4.5.4 Explicit 4 × 4 structure

The explicit form of the effective field tensor is

$$G_{\mu\nu}^{(\mathrm{eff})} = \begin{pmatrix} 0 & -g_1^{(\mathrm{eff})}/c & -g_2^{(\mathrm{eff})}/c & -g_3^{(\mathrm{eff})}/c \\ +g_1^{(\mathrm{eff})}/c & 0 & +b_3 & -b_2 \\ +g_2^{(\mathrm{eff})}/c & -b_3 & 0 & +b_1 \\ +g_3^{(\mathrm{eff})}/c & +b_2 & -b_1 & 0 \end{pmatrix}$$

with

$$\mathbf{g}^{(\mathrm{eff})} = \mathbf{g} + \nabla\left(\frac{GM\,h^2}{c^2 r^3}\right)$$

$$\mathbf{b}^{(\mathrm{eff})} = \mathbf{b}$$

4.5.5 Interpretation

The orbital correction modifies only the time–space sector of the field tensor. This structure mirrors the electromagnetic case where the addition of a scalar potential leaves the magnetic field unchanged, adding a gradient to the electric field. In the present gravitational context, the term

$$\phi^{(\mathrm{orb})} = -\frac{GM\,h^2}{c^2 r^3}$$

introduces a kinematical correction to the radial component of the field, which will feed directly into the effective acceleration in the next subsection.

4.6 Effective Equation of Motion

4.6.1 The axiomatic dynamical law

The axiomatic dynamical law of the flat-spacetime gravitational theory in the presence of the effective field is taken to be

$$a^\mu = G_\nu^{(\mathrm{eff})\mu}\, U^\nu$$

where U^μ is the four–velocity of the test body and $G_{\mu\nu}^{(\mathrm{eff})}$ the antisymmetric effective field tensor constructed from $\Phi^{(\mathrm{eff})\mu}$. This is the exact gravitational analogue of the electromagnetic relation

$$\frac{q}{m}\, F_\nu^\mu U^\nu$$

with the crucial difference that here the coupling constant is absorbed into $G_{\mu\nu}^{(\mathrm{eff})}$ itself: the four–acceleration is directly equal to the effective field acting on the four–velocity.

4.6.2 Orthogonality of a^μ and U^μ

Contracting the equation of motion with U_μ gives

$$U_\mu a^\mu = U_\mu G_\nu^{(\mathrm{eff})\mu} U^\nu = 0$$

because $G_{\mu\nu}^{(\mathrm{eff})}$ is antisymmetric. Thus

$$U_\mu U^\mu = -c^2 \quad \Rightarrow \quad \frac{d}{d\tau}\left(U_\mu U^\mu\right) = 0$$

and the mass–shell condition is preserved along the worldline. The effective field changes the direction of U^μ but never its invariant magnitude.

4.6.3 Three–dimensional form: Coulomb–Lorentz structure

Using the relations

$$G_{0j}^{(\text{eff})} = -\frac{g_i^{(\text{eff})}}{c}$$
$$G_{ij}^{(\text{eff})} = \varepsilon_{ijk} b_k$$

and writing $U^\mu = (U^0, \mathbf{u})$, the spatial components of the equation of motion become

$$a^i = G_\nu^{(\text{eff})i} U^\nu = G_0^{(\text{eff})i} U^0 + G_j^{(\text{eff})i} U^j$$

In the rest frame of the source, $G_0^{(\text{eff})i} = g_i^{(\text{eff})}/c$ and $G_j^{(\text{eff})i} = \varepsilon_{ijk} b_k$, so that

$$\mathbf{a} = \gamma \mathbf{g}^{(\text{eff})} + \mathbf{u} \times \mathbf{b}$$

so, since $\mathbf{u} = \gamma \mathbf{v}$ and

$$\mathbf{a} = \frac{\mathrm{d}\mathbf{u}}{d\tau} = \gamma \frac{\mathrm{d}\mathbf{u}}{dt}$$

It follows that:

$$\frac{\mathrm{d}\mathbf{u}}{dt} = \mathbf{g}^{(\text{eff})} + \mathbf{v} \times \mathbf{b}$$

The spatial acceleration thus has a pure "Coulomb'' part $\mathbf{g}^{(\text{eff})}$ and a "Lorentz''–type part $\mathbf{v} \times \mathbf{b}$, completing the analogy with the electromagnetic force, where the electric and magnetic contributions appear in exactly the same vector combination.

4.6.4 Static sources: the case $b = 0$

For a static, spherically symmetric source, the gravito-rotational field vanishes, $\mathbf{b} = \mathbf{0}$, and the effective acceleration reduces to

$$\frac{d\mathbf{u}}{dt} = \mathbf{g}^{(\text{eff})}$$
$$\mathbf{g}^{(\text{eff})} = \mathbf{g} - \nabla\phi^{(\text{orb})}$$

In this case the motion is governed entirely by a generalized Coulomb–type field, with the orbital correction encoded in the scalar potential $\phi^{(\text{orb})}$.

4.6.5 Spherical symmetry and the radial component

For a static, spherically symmetric source of mass M,

$$\mathbf{g} = -\frac{GM}{r^2}\hat{\mathbf{r}}, \qquad \phi^{(\text{orb})} = -\frac{GM\,h^2}{c^2 r^3}$$

so that

$$\frac{d\mathbf{u}}{dt} = -\frac{GM}{r^2}\hat{\mathbf{r}} - \nabla\left(-\frac{GM\,h^2}{c^2 r^3}\right) \Rightarrow$$

$$\frac{d\mathbf{u}}{dt} = -\frac{GM}{r^2}\hat{\mathbf{r}} - \frac{3GM\,h^2}{c^2 r^4}\hat{\mathbf{r}}$$

Consequently, the radial component of the spatial acceleration is

$$\frac{d\mathbf{u}}{dt} = a_r\hat{\mathbf{r}}$$
$$a_r = -\frac{GM}{r^2} - \frac{3GM\,h^2}{c^2 r^4}$$

reproducing the familiar post–Newtonian correction responsible for phenomena such as the perihelion advance of bounded orbits.

4.6.6 Temporal component and power

The temporal component of the equation of motion is

$$a^0 = G_\nu^{(\text{eff})0} U^\nu = \frac{1}{c}\, \mathbf{g}^{(\text{eff})} \cdot \mathbf{u}$$

which governs the rate of change of $U^0 = \gamma c$ with proper time:

$$\frac{dU^0}{d\tau} = \frac{1}{c}\, \mathbf{g}^{(\text{eff})} \cdot \mathbf{u}$$

In terms of the coordinate time t, this gives the power per unit mass transferred by the effective field to the particle:

$$\frac{d}{dt}(\gamma c^2) = \mathbf{g}^{(\text{eff})} \cdot \mathbf{v}$$

and multiplying by the rest mass m_0

$$\frac{d}{dt}(mc^2) = \mathrm{m}_0\, \mathbf{g}^{(\text{eff})} \cdot \mathbf{v}$$

showing that the scalar product $\mathbf{g}^{(\text{eff})} \cdot \mathbf{v}$ plays the role of an effective power density in the Flat Spacetime Gravity framework.

4.6.7 A fully relativistic force framework in flat spacetime

Once the fundamental dynamical law

$$\alpha^\mu = G_\nu^{(\text{eff})\mu}\, U^\nu$$

has been established, an important structural fact becomes visible. The kinematical tensor W_ν^μ appearing in $\alpha^\mu = W_\nu^\mu\, U^\nu$ acts as a *universal receptacle*: any additional force field that couples linearly to the 4–velocity is simply absorbed as an additive contribution inside the same operator. For instance, in the simultaneous presence of an electromagnetic field,

$$m_0\, a^\mu \;=\; m_0\, G_\nu^{(\mathrm{eff})\mu} U^\nu \;+\; q_0\, F_\nu^\mu U^\nu$$

and therefore, dividing by m_0,

$$W_\nu^\mu\, U^\nu \;=\; G_\nu^{(\mathrm{eff})\mu} U^\nu \;+\; \frac{q_0}{m_0}\, F_\nu^\mu U^\nu$$

Thus, the same W–tensor that governs pure gravitational motion automatically incorporates the electromagnetic acceleration through the factor q_0/m_0. The dynamical content is not modified: only the "source" of the acceleration changes.

This additive structure reflects a deeper aspect of the Flat Spacetime Gravity formulation: the theory provides a strictly relativistic, flat–spacetime framework in which *all* long–range force fields can be treated on the same footing. Instead of curving spacetime, Flat Spacetime Gravity encodes dynamics in the antisymmetric tensor operators acting on U^μ. Gravitational, electromagnetic (and, in principle, any other field with a 2–form field strength) contributions enter the motion equation in a single, transparent way, as linear terms inside W_ν^μ. This unification—while entirely kinematical—is remarkably powerful: a single relativistic operator governs the response of matter to all fields present, and the structure remains strictly Minkowskian throughout.

4.6.8 Final remarks

The relation

$$a^\mu \;=\; G_\nu^{(\mathrm{eff})\mu}\, U^\nu$$

encapsulates the entire dynamics of a test body in the effective flat-spacetime gravitational field. Its structure is the exact gravitational analogue of the Lorentz force in electromagnetism, with the sourceful and gravito-rotational fields $(\mathbf{g}^{(\mathrm{eff})}, \mathbf{b})$ playing the roles of electric and magnetic components. At the same time, the antisymmetry of $G_{\mu\nu}^{(\mathrm{eff})}$ guarantees the orthogonality of a^μ and U^μ and the preservation of the mass–shell condition, while the explicit 3D form reproduces both the

Newtonian limit and the standard post–Newtonian corrections within a strictly Minkowskian spacetime.

4.7 Conservation Theorems in the Effective Field

4.7.1 Conservation of energy, Generalized Momentum and Angular Momentum

The effective dynamical law

$$a^{\mu} = G_{\nu}^{(\text{eff})\mu}\, U^{\nu}$$

together with the antisymmetry of $G_{\mu\nu}^{(\text{eff})}$ implies three fundamental conservation theorems. These arise directly from the equation of motion and do not require a variational principle.

4.7.2 Conservation of Energy

From the temporal component of the equation of motion,

$$\mathrm{a}^{0} = G_{\nu}^{(\text{eff})0}\, \mathrm{U}^{\nu} = \frac{1}{\mathrm{c}}\, \mathbf{g}^{(\text{eff})} \cdot \mathbf{u}$$

and using $U^{0} = \gamma c$, we obtain

$$\frac{\mathrm{d}(\gamma\mathrm{c})}{\mathrm{d}\tau} = \frac{1}{\mathrm{c}}\, \mathbf{g}^{(\text{eff})} \cdot \mathbf{u}$$

For static fields, $\mathbf{g}^{(\text{eff})} = -\nabla\phi^{(\text{eff})}$, and thus,

$$\frac{\mathrm{d}(\gamma\mathrm{c})}{\mathrm{d}\tau} = -\frac{1}{\mathrm{c}}\nabla\phi^{(\text{eff})} \cdot \mathbf{u}$$

$$\frac{\mathrm{d}(\gamma\mathrm{c})}{\mathrm{d}\tau} = -\frac{1}{\mathrm{c}}\frac{\mathrm{d}\phi^{(\text{eff})}}{\mathrm{d}\tau}$$

Hence

$$d\gamma = -\frac{d\phi^{(\mathrm{eff})}}{c^2}$$

so

$$\gamma = 1 - \frac{\phi^{(\mathrm{eff})}}{c^2} + \frac{E}{c^2}$$

where *E* is the conserved specific energy (energy per unit mass).

Multiplying by $m_0 c^2$ gives

$$m_0\, E = mc^2 - m_0 c^2 + m_0 \phi^{(\mathrm{eff})}, \qquad m = m_0 \gamma$$

Newtonian orbital limit

In the nonrelativistic limit $\gamma \approx 1 + 1/2\, v^2/c^2$,

$$1 + \frac{v^2}{2c^2} \approx 1 - \frac{\phi^{(\mathrm{eff})}}{c^2} + \frac{E}{c^2} \quad \Rightarrow$$

$$v^2 = -2\,\phi^{(\mathrm{eff})} + 2\,E$$

which is the Newtonian energy relation,

$$\frac{v^2}{2} + \phi^{(\mathrm{eff})} = E$$

Black–hole condition

Requiring that the escape velocity satisfy $v_{\mathrm{esc}} \geq c$ gives

$$c^2 \leq v_{\mathrm{esc}}^2 = -2\,\phi^{(\mathrm{eff})} + 2\,E$$

For photons, $\phi^{(\mathrm{orb})} = 0$ and $E = 0$, hence

$$-2\phi \geq c^2 \quad \Rightarrow \quad R = \frac{2GM}{c^2}$$

Thus, the familiar Schwarzschild radius emerges directly in Minkowski space from the effective potential.

4.7.3 Conservation of Generalized Momentum

Start from the covariant dynamical law

$$\frac{dU_\mu}{d\tau} = G_{\mu\nu}^{(\mathrm{eff})} U^\nu, \qquad G_{\mu\nu}^{(\mathrm{eff})} = \partial_\mu \Phi_\nu^{(\mathrm{eff})} - \partial_\nu \Phi_\mu^{(\mathrm{eff})}$$

Insert the explicit form of $G_{\mu\nu}^{(\mathrm{eff})}$:

$$\frac{dU_\mu}{d\tau} = \left(\partial_\mu \Phi_\nu^{(\mathrm{eff})}\right) U^\nu - \left(\partial_\nu \Phi_\mu^{(\mathrm{eff})}\right) U^\nu$$

Rearrange terms by grouping derivatives of $\Phi_\mu^{(\mathrm{eff})}$:

$$\frac{dU_\mu}{d\tau} + \left(\partial_\nu \Phi_\mu^{(\mathrm{eff})}\right) U^\nu = \left(\partial_\mu \Phi_\nu^{(\mathrm{eff})}\right) U^\nu$$

Observe that the left–hand side is exactly the total proper–time derivative of $U_\mu + \Phi_\mu^{(\mathrm{eff})}$:

$$\frac{d}{d\tau}\left(U_\mu + \Phi_\mu^{(\mathrm{eff})}\right) = \left(\partial_\mu \Phi_\nu^{(\mathrm{eff})}\right) U^\nu - \left(\partial_\nu \Phi_\mu^{(\mathrm{eff})}\right) U^\nu$$

Now impose that the effective potential is *homogeneous* in the coordinate x^α, meaning

$$\partial_\alpha \Phi_\nu^{(\mathrm{eff})} = 0 \qquad \text{and} \qquad \partial_\alpha \Phi_\mu^{(\mathrm{eff})} = 0$$

Evaluating the previous expression for $\mu = \alpha$ gives

$$\frac{d}{d\tau}\left(U_\alpha + \Phi_\alpha^{(\mathrm{eff})}\right) = 0$$

Hence the quantity of generalized momentum per unit mass

$$\Pi_\mu^{(\mathrm{eff})} = \left(U_\mu + \Phi_\mu^{(\mathrm{eff})}\right)$$

and the generalized momentum

$$P_\mu^{(\mathrm{eff})} = m_0 \left(U_\mu + \Phi_\mu^{(\mathrm{eff})}\right)$$

is conserved along the worldline whenever the effective potential is independent of the coordinate x^{α}.

4.7.4 Conservation of Angular Momentum

The orbital bivector is

$$H_{\mu\nu} = X_{\mu} U_{\nu} - X_{\nu} U_{\mu} = X \wedge U$$

Differentiation along the worldline gives

$$\frac{dH_{\mu\nu}}{d\tau} = \frac{dX_{\mu}}{d\tau} U_{\nu} + X_{\mu}\alpha_{\nu} - X_{\nu}\alpha_{\mu} = X_{\mu}\alpha_{\nu} - X_{\nu}\alpha_{\mu} = X \wedge \alpha$$

Hence,

$$\frac{dH_{\mu\nu}}{d\tau} = 0 \quad \Leftrightarrow \quad X \wedge \alpha = 0$$

In spatial form this becomes

$$\frac{d\mathbf{h}}{d\tau} = \mathbf{r} \times \mathbf{a}$$

If the effective acceleration is central,

$$\mathbf{a} = \gamma\, \mathbf{g}^{(\mathrm{eff})} + \mathbf{u} \times \mathbf{b} \quad \text{with} \quad \mathbf{g}^{(\mathrm{eff})} \parallel \mathbf{r}$$

then

$$\mathbf{r} \times \mathbf{a} = \mathbf{0} \quad \Rightarrow \quad \frac{d\mathbf{h}}{d\tau} = 0$$

Thus, the spatial angular momentum **h** is conserved for all motions in a spherically symmetric effective field.

4.8 Relativistic Two–Body Structure in Flat Spacetime Gravity

4.8.1 Introductory remark

In Newtonian mechanics the two–body problem admits a clean reduction: the interaction force depends only on the separation vector, the total momentum is conserved, and the dynamics separates into a center–of–mass motion and an effective one–body problem. In relativistic field theories, however, this separation is no longer transparent, because the field itself carries energy–momentum and the particle momentum is not simply $m\mathbf{v}$.

As we saw, from the generalized momentum per unit mass

$$\Pi_{\mu}^{(\text{eff})} = \left(\mathrm{U}_{\mu} + \Phi_{\mu}^{(\text{eff})}\right)$$

we can construct the generalized momentum

$$\mathrm{P}_{\mu}^{(\text{eff})} = \mathrm{m}_0 \left(\mathrm{U}_{\mu} + \Phi_{\mu}^{(\text{eff})}\right)$$

which plays a far more profound role in Flat Spacetime Gravity than a mere formal analogue of the electromagnetic minimal coupling. As we shall demonstrate below, the spatial part

$$\mathbf{P}_{\text{NR}} = m_0\mathbf{v} + m_0\mathbf{A}^{(\text{eff})}$$

encodes a *hidden two–body structure*: while $m_0\mathbf{v}$ represents the ordinary probe momentum, the term $m_0\mathbf{A}^{(\text{eff})}$ captures — in a fully relativistic way — the kinematic imprint of the source. This mirrors the role played by the reduced mass in the Newtonian two–body problem, but now elevated to a clean Minkowskian formulation directly tied to the gravitational potentials $(\phi^{(\text{eff})}, \mathbf{A}^{(\text{eff})})$. Thus, the generalized momentum becomes the natural carrier of the probe–source dynamical coupling in Flat Spacetime Gravity, and forms the correct conserved quantity whenever the background exhibits the corresponding symmetry.

4.8.2 The generalized momentum as a two–body object

For a probe particle of rest mass m_0 moving in a gravitational effective potential $\Phi^{(\mathrm{eff})\,\mu}$ generated by a source of rest mass M_0, the generalized momentum is

$$\mathrm{P}_\mu^{(\mathrm{eff})} = m_0 U_\mu + m_0 \Phi_\mu^{(\mathrm{eff})}$$

The first term is the kinematical momentum of the probe. The second term is subtler: using the relation

$$m_0\,\Phi^{(\mathrm{eff})} = -G\,\frac{m_0 M_0}{r} - G\,\frac{m_0 M_0 h^2}{c^2 r^3}$$

we see that $m_0\,\Phi^{(\mathrm{eff})}$ plays the role of the interaction momentum that normally appears inside the total momentum

$$P^\mu_{\mathrm{total}} = m_0 U^\mu + M_0 V^\mu + (\text{field momentum})$$

Thus, the combination

$$\mathrm{P}^{(\mathrm{eff})\mu} = \mathrm{m}_0(\mathrm{U}^\mu + \Phi^{(\mathrm{eff})\mu})$$

acts as a compressed, covariant representation of the *two–body momentum* of the full system.

4.8.3 Covariant two–body reduction in Flat Spacetime Gravity

Although relativity has no "reduced mass" in the Newtonian sense, Flat Spacetime Gravity produces a structural analogue:

$$P^\mu_{\mathrm{total}} = P^\mu_{\mathrm{probe}} + P^\mu_{\mathrm{source}} + P^\mu_{\mathrm{field}} \quad \leftrightarrow \quad m_0 U^\mu + m_0 \Phi^{(\mathrm{eff})\mu}$$

Here $m_0\Phi^{(\mathrm{eff})\,\mu}$ represents the momentum transferred from the source to the probe through the field, so that the generalized momentum of the probe already includes the two–body structure of the system.

That the two–body structure appears *inside* the particle-level generalized momentum is a highly nontrivial feature. It signals that Flat Spacetime Gravity naturally incorporates matter–field momentum exchange and unifies particle and field dynamics at the level of variational principles.

4.8.4 A 3–dimensional nonrelativistic limit

To connect the covariant combination

$$\mathrm{p}^{(\mathrm{eff})\mu} = m_0\left(U^\mu + \Phi^{(\mathrm{eff})\mu}\right)$$

with the familiar Newtonian structures, let us consider the nonrelativistic regime, where $|\mathbf{v}| \ll c$ and the Lorentz factor reduces to

$$\gamma = \frac{1}{\sqrt{1-\mathbf{v}^2/c^2}} \approx 1+\frac{\mathbf{v}^2}{2c^2}$$

In Flat Spacetime Gravity the four–velocity and the gravitational effective four–potential of the source are written as

$$U^\mu = \gamma\,(c,\mathbf{v})$$

$$\Phi^{(\mathrm{eff})} = \left(\frac{\phi^{(\mathrm{eff})}}{c},\mathbf{A}^{(\mathrm{eff})}\right) = \left(\frac{\phi}{c},\mathbf{A}\right) + \left(\frac{\phi^{(\mathrm{orb})}}{c},\mathbf{A}^{(\mathrm{orb})}\right)$$

where $\phi^{(\mathrm{eff})}$ is the effective scalar potential and $\mathbf{A}^{(\mathrm{eff})}$ the effective vector potential.

Temporal component

The temporal component of the generalized momentum reads

$$\mathrm{p}^{(\mathrm{eff})0} = m_0\left(U^0 + \Phi^{(\mathrm{eff})0}\right) = m_0\left(\gamma c + \frac{\phi^{(\mathrm{eff})}}{c}\right)$$

Using the expansion for γ,

$$\begin{aligned}\mathrm{p}^{(\mathrm{eff})0} &\approx m_0\left[c\left(1+\frac{\mathbf{v}^2}{2c^2}\right)+\frac{\phi^{(\mathrm{eff})}}{c}\right]\\ &= \frac{m_0c^2}{c} + \frac{\phi^{(\mathrm{eff})}}{c} + \frac{1}{2}\frac{m_0\mathbf{v}^2}{c}\end{aligned}$$

Multiplying by c we obtain the energy–like combination

$$c\,\mathrm{P}^{(\mathrm{eff})0} \approx m_0c^2 + m_0\phi^{(\mathrm{eff})} + \frac{1}{2}m_0\mathbf{v}^2$$

that is, rest energy plus gravitational potential energy plus kinetic energy, exactly as in Newtonian mechanics.

Spatial component

For the spatial part we have

$$\mathbf{P}^{\mathrm{eff}} = m_0\left(\mathbf{U} + \mathbf{\Phi}^{(\mathrm{eff})}\right) = m_0\left(\gamma\,\mathbf{v} + \mathbf{A}^{(\mathrm{eff})}\right)$$

In the nonrelativistic limit $\gamma \approx 1$, hence

$$\mathbf{P}^{\mathrm{eff}} \approx m_0\mathbf{v} + m_0\mathbf{A}^{(\mathrm{eff})} = m_0\mathbf{v} + m_0\mathbf{A}$$

The first term is the ordinary Newtonian momentum of the probe; the second term behaves as an effective momentum contribution originating from the source through the gravitational potential. Thus, in three dimensions,

$$c\,\mathrm{P}^0_{\mathrm{NR}} \approx m_0c^2 + m_0\phi^{(\mathrm{eff})} + \frac{1}{2}m_0\mathbf{v}^2$$

$$\mathbf{P}_{\mathrm{NR}} \approx m_0\mathbf{v} + m_0\mathbf{A}$$

This shows that the covariant FSG combination $m_0(U^\mu + \Phi^{(\mathrm{eff})\,\mu})$ reduces, in the nonrelativistic limit, to the usual Newtonian momentum $m_0\mathbf{v}$ plus a field–induced contribution $m_0\mathbf{A}$, and to the standard decomposition of the energy into rest, potential, and kinetic parts. The two–body character of the relativistic object $\mathrm{P}^{(\mathrm{eff})\,\mu}$ is therefore consistent with the familiar Newtonian two–body structures in the appropriate limit.

4.8.5 Decomposition of the spatial momentum and the two–body structure

In the nonrelativistic limit we found that

$$\mathbf{P}_{\mathrm{NR}} = m_0\mathbf{v} + m_0\mathbf{A}$$

We now explain in detail why the second term $m_0\mathbf{A}$ is not an arbitrary correction, but the *flat–spacetime analogue of the two–body momentum redistribution* familiar from Newtonian mechanics.

The vector potential as the kinematic imprint of the source

In Flat Spacetime Gravity the gravitational effective vector potential of the source is generated from its own four–velocity V^{μ} according to

$$\Phi^{(\mathrm{eff})\mu} = \left(\frac{\phi^{(\mathrm{eff})}}{c}, \mathbf{A}^{(\mathrm{eff})}\right)$$

$$\mathbf{A}^{(\mathrm{eff})} = \mathbf{A} \propto \frac{GM}{c^2}\, \mathbf{v}_{\mathrm{source}} \times \frac{\mathbf{r}}{r^3} \quad \text{(orbital part)}$$

Thus $\mathbf{A}$ encodes the *orbital motion of the source mass*, exactly as in electromagnetism the magnetic vector potential encodes the motion of the charge distribution. The term $m_0\mathbf{A}$ is therefore proportional to the source velocity $\mathbf{v}_{\mathrm{source}}$ and acts as an effective momentum transfer between the two bodies.

Hidden reduced–mass structure

In Newtonian mechanics, the two–body problem reorganizes into a center–of–mass motion plus a relative motion:

$$m\dot{\mathbf{r}} \;\rightarrow\; \mu\dot{\mathbf{r}}_{\mathrm{rel}}, \qquad \mu = \frac{mM}{m+M}$$

In Flat Spacetime Gravity, the relativistic object

$$\mathrm{P}^{(\mathrm{eff})\mu} = m_0(U^{\mu} + \Phi^{(\mathrm{eff})\mu})$$

plays the role of the *relativistic analogue of the reduced–mass momentum*. The term $m_0\Phi^{(\mathrm{eff})\mu}$ injects the kinematics of the heavy source into the momentum of the light probe — precisely the relativistic generalization of how the reduced mass absorbs both masses into a single effective momentum.

Hence the spatial part

$$\mathbf{P}_{\mathrm{NR}} = m_0\mathbf{v} + m_0\mathbf{A}$$

should be interpreted as

$$\mathbf{P}_{\mathrm{NR}} = (\text{probe momentum}) + (\text{source-encoded momentum})$$

This is why $P^{(\mathrm{eff})\,\mu}$ is conserved when the spacetime is homogeneous along the chosen coordinate: it is the *total two–body momentum* of probe+source in flat spacetime geometry.

Summary

- The spatial momentum $m_0\mathbf{v}$ is the usual probe momentum.
- The additional term $m_0\mathbf{A}$ is the relativistic imprint of the source motion on the probe, exactly analogous to the role of reduced mass.
- The orbital potential $\phi^{(\mathrm{orb})}$ and the vector potential $\mathbf{A}$ describe the same physics from different angles: the two–body orbital coupling in Flat Spacetime Gravity.
- The generalized momentum $P^{(\mathrm{eff})\mu} = m_0(U^\mu + \Phi^{(\mathrm{eff})\mu})$ is therefore the natural relativistic two–body momentum in flat spacetime geometry.

This clarifies why the Flat Spacetime Gravity formulation preserves total momentum in the expected way, and embeds the Newtonian two–body problem into a clean Minkowski–space formalism.

4.8.6 Orbital Potential from the Coupling of h and b

The heuristic relation below should be read as structural consistency, while the exact result is fixed by the 4D tensorial derivation, as in the previous subsections.

A key structural feature of Flat Spacetime Gravity is that the *effective* gravitational potential experienced by a probe consists of two contributions:

$$\phi^{(\mathrm{eff})} = \phi + \phi^{(\mathrm{orb})}$$

$$\phi = -\frac{GM}{r}, \qquad \phi^{(\mathrm{orb})} = -\frac{GMh^2}{c^2 r^3}$$

The first term, $\phi = -GM/r$, is the usual source–centered potential. The second term, $\phi^{(\mathrm{orb})}$, is the characteristic FSG correction. We now show that this correction arises naturally and unavoidably from the intrinsic coupling between:

- the probe's orbital angular momentum

$$\mathbf{h} = \mathbf{r} \times \mathbf{u}, \qquad h = |\mathbf{h}|$$

- the *rotational flat–spacetime gravitational field* of the moving source,

$$\mathbf{b} = \nabla \times \mathbf{A}$$

For a moving mass, the vector potential **A** takes the universal gravitational Liénard–Wiechert form (nonrelativistic source for clarity):

$$\mathbf{A} = -\frac{GM}{c^2}\frac{\mathbf{v}_{\mathrm{source}} \times \mathbf{r}}{r^3}$$

so that

$$\mathbf{b} = \nabla \times \mathbf{A} = -\frac{GM}{c^2}\left(3\frac{(\mathbf{v}_{\mathrm{source}} \cdot \mathbf{r})\,\mathbf{r}}{r^5} - \frac{\mathbf{v}_{\mathrm{source}}}{r^3}\right)$$

If the source executes orbital motion about the center of mass, its velocity is orthogonal to **r**:

$$\mathbf{v}_{\mathrm{source}} \cdot \mathbf{r} = 0$$

and the expression collapses to

$$\mathbf{b} = +\frac{GM}{c^2}\frac{\mathbf{v}_{\mathrm{source}}}{r^3}$$

For a two–body system, Newtonian kinematics ensures that both bodies share the same orbital angular velocity:

$$\boldsymbol{\omega}_{\text{orb}} = \frac{\mathbf{h}}{r^2}$$

$$\mathbf{v}_{\text{source}} = \boldsymbol{\omega}_{\text{orb}} \times \mathbf{r}$$

Thus,

$$\mathbf{b} \;=\; +\,\frac{GM}{c^2}\frac{\boldsymbol{\omega}_{\text{orb}} \times \mathbf{r}}{r^3} \;=\; +\,\frac{GM}{c^2}\frac{\mathbf{h}}{r^3}$$

Consequently, the unique scalar that can be formed from **h** and **b** is

$$\mathbf{h}\cdot\mathbf{b} \;=+\,\frac{GM}{c^2}\frac{h^2}{r^3}$$

Because **h** is a constant of motion and **b** is divergence–free, the combination $\mathbf{h}\cdot\mathbf{b}$ is the *only* rotationally invariant scalar available in the problem. Dimensional analysis then requires the orbital correction to be

$$\phi^{(\text{orb})} \;\propto\; -\,\mathbf{h}\cdot\mathbf{b} \;=\; -\,\frac{GM}{c^2}\frac{h^2}{r^3}$$

and substituting the above expression gives the exact FSG value:

$$\phi^{(\text{orb})}(r) \;=\; -\,\frac{GMh^2}{c^2 r^3}$$

Thus, the celebrated orbital term of Flat Spacetime Gravity is neither an ad hoc postulate nor an imported analogy from electromagnetism. It follows inevitably from the unique scalar coupling between the probe's orbital angular momentum and the rotational gravitational field of the moving source.

4.8.7 Closing Remark

In the same way as the acceleration tensor $W_{\mu\nu}$ decomposes into purely geometric components: orbital transport, Thomas–Fermi precession, and spin–induced rotation, a parallel structural rigidity appears when we analyze the generalized momentum $P^{(\mathrm{eff})\mu} = m_0 U^{\mu} + m_0 \Phi^{(\mathrm{eff})\mu}$. Its conservation properties reveal that the familiar Newtonian "two–body reduction'' (centre–of–mass motion plus relative motion) re-emerges in Flat Spacetime Gravity as a fully covariant statement: the source–encoded potential terms behave exactly like the relativistic momentum of the companion body. Thus, the effective potential is not an auxiliary construct but the covariant bookkeeping device that carries the dynamical influence of both bodies while maintaining the strict flatness of FSG geometry.

The culmination is the scalar identity

$$\phi^{(\mathrm{orb})}(r) = -\frac{GMh^2}{c^2 r^3}$$

extracted from the unique rotational invariant $\mathbf{h} \cdot \mathbf{b}$. This shows that the celebrated orbital term of Flat Spacetime Gravity is not an ad hoc postulate but an unavoidable consequence of the gravitational Liénard–Wiechert structure of a moving source. The correction appears exactly where and only where the geometry allows a rotational scalar to be formed, and its magnitude is then dictated by dimensional analysis and by the dynamical invariants of the two–body system. No curvature, no metric deformation, and no hidden assumptions are required.

Taken together, these results elevate this section from a merely technical development to a conceptual cornerstone. It exhibits, in its most transparent form, how flat–spacetime gravitation can encode both the kinematics *and* the dynamics of a multi–body system, while producing all the effective structures that in curved–spacetime theories are often attributed to spacetime geometry itself. In this sense, this section prepares the conceptual ground for the quantum developments that follow: once the covariant momentum and rotation structures are understood at the classical level, their canonical and quantum counterparts emerge with remarkable clarity.

4.9 Concluding Remarks

The developments of this section establish a complete and internally consistent relativistic dynamical framework for the Flat Spacetime Gravity. All gravitational dynamics are encoded in the antisymmetric tensor $G_{\mu\nu}$ and in the kinematical bivector $H_{\mu\nu}$, without any reference to curved geometry. A key element is the construction of the *orbital potential* $\Phi^{(\mathrm{orb})\mu}$, which provides the unique Lorentz–covariant coupling between the source field $G_{\mu\nu}$ and the intrinsic Fermi motion of the test body. This yields the *effective potential* $\Phi^{(\mathrm{eff})\mu}$ and its field tensor $G_{\mu\nu}^{(\mathrm{eff})}$, so that the flat–space dynamical law takes the form

$$a^{\mu} \;=\; G_{\nu}^{(\mathrm{eff})\mu}\, U^{\nu}$$

a perfect gravitational analogue of the electromagnetic Lorentz force.

The spatial projection of this relation cleanly separates into an *ergic* (work–producing) part and a rotational Lorentz–type part, thereby reproducing—in a fully relativistic but flat–space way—the Newtonian force law plus the characteristic post–Newtonian corrections. The perihelion shift, light deflection, and the escape–velocity condition leading to the Schwarzschild radius all emerge from $G^{(\mathrm{eff})}$ alone, without invoking curvature, weak–field expansions, or perturbative metrics.

At a deeper level, this section reveals a structural feature normally hidden in standard approaches. The combination

$$\mathrm{P}_{\mu}^{(\mathrm{eff})} \;=\; \mathrm{m}_0\left(\mathrm{U}_{\mu} + \Phi_{\mu}^{(\mathrm{eff})}\right)$$

acts as the *conserved generalized momentum* of the entire two–body system (probe + source), with the orbital part $\Phi^{(\mathrm{orb})\mu}$ encoding the influence of the source motion in exactly the same way that the vector potential encodes electromagnetic source motion. The relativistic two–body structure becomes explicit: U^{μ} represents the probe motion, while $m_0\Phi^{(\mathrm{eff})\mu}$ plays the role of the source–encoded contribution. The orbital potential therefore functions as a relativistic generalization of reduced–mass dynamics, embedded directly into Minkowski spacetime.

A striking consequence is the natural appearance of the scalar combination $(\mathbf{h} \cdot \mathbf{b})$, where $\mathbf{h}$ is the orbital angular momentum and $\mathbf{b}$ the rotational flat–spacetime field. This coupling,

$$\Phi^{(\mathrm{orb})} \propto -\mathbf{h} \cdot \mathbf{b} \sim -\frac{GMh^2}{c^2 r^3}$$

is precisely the term responsible for the relativistic corrections to orbital motion. Its appearance here is not postulated but *deduced* from the intrinsic geometry of the antisymmetric sector of the theory. In this way Flat Spacetime Gravity reproduces the known phenomenology while also offering a transparent mechanical interpretation: the field $\mathbf{b}$ couples directly to the transported orbital angular momentum of the probe.

In conclusion, this section establishes that Flat Spacetime Gravity admits a fully relativistic, Lorentz–covariant gravitational dynamics, complete conservation laws, and a natural two–body structure, all without curvature. It prepares the ground for the Lagrangian and Hamiltonian formulations of the next sections and reveals that the gravitational corrections usually attributed to spacetime curvature arise here from a purely Minkowskian antisymmetric field geometry.

5. Astronomical tests

5.1 Introductory Remark

The effective dynamics developed in previous sections provides a unified expression for the acceleration of a test body in a spherically symmetric field,

$$\frac{d\boldsymbol{u}}{dt} = -\left(\frac{GM}{r^2} + \frac{3GMh^2}{c^2r^4}\right)\boldsymbol{e}_r$$

obtained entirely within Minkowski spacetime through the combination of the source potential and the orbital potential. This last ingredient, $\Phi^{(\mathrm{orb})\mu}$, is the crucial innovation: it introduces a natural coupling between the external field and the trajectory of the particle itself, producing a dynamical correction that has no analogue in Newtonian gravity and does not arise from Maxwell–type equations by itself. Its presence modifies the radial force law in a way that reproduces the familiar post–Newtonian effects without invoking spacetime curvature.

The purpose of the present chapter is to examine the observational consequences of this effective acceleration. Starting from the same dynamical expression, we derive the four classical astronomical tests traditionally associated with general relativity: the perihelion advance of bound orbits, the deflection of light, the gravitational redshift of frequencies, and the Shapiro time delay of electromagnetic signals. Each test follows directly from the flat–space dynamics of the effective field and requires no geometric deformation of the underlying spacetime.

In this way, the astronomical predictions of the flat-spacetime gravitational theory are placed on the same empirical footing as those of general relativity, while arising from a linear, Maxwell–like structure rather than from curved geometry.

In particular, light deflection will be shown to emerge from the velocity-independent orbital potential, while the source-generated contribution is dynamically suppressed in the null limit through the relativistic growth of the proper-velocity **u**.

5.2 Perihelion Precession

For a static, spherically symmetric source, as shown in the previous section, the effective three-dimensional dynamical law reduces to

$$\frac{d\mathbf{u}}{dt} = -a_r \mathbf{e}_r$$

with the radial component of the acceleration

$$a_r = -\left(\frac{GM}{r^2} + \frac{3GMh^2}{c^2 r^4}\right)$$

Here h is the conserved specific angular momentum of the orbit, as defined in the previous sections.

To study the orbital shape, it is convenient to introduce the standard Binet variable

$$u(\theta) = \frac{1}{r(\theta)}$$

where θ is the polar angle in the orbital plane. For planar motion with conserved h, the radial acceleration can be written as

$$a_r = -h^2 u^2 \left(\frac{d^2 u}{d\theta^2} + u\right)$$

Equating this with the effective radial acceleration gives the Binet-type equation

$$-h^2 u^2 \left(\frac{d^2 u}{d\theta^2} + u\right) = -\left(GMu^2 + 3GM\,\frac{h^2}{c^2}\,u^4\right)$$

or, after division by $-u^2$,

$$h^2 \left(\frac{d^2 u}{d\theta^2} + u\right) = GM + 3GM\,\frac{h^2}{c^2}\,u^2$$

To first order in GM/c^2 we write $u = u_0 + u_1$, where u_0 solves the Newtonian equation

$$\frac{d^2u_0}{d\theta^2} + u_0 = \frac{GM}{h^2}$$

with solution

$$u_0(\theta) = \frac{GM}{h^2}(1 + e\cos\theta)$$

e being the eccentricity of the orbit.

Substituting $u = u_0 + u_1$ into the full equation and retaining only terms linear in GM/c^2 yields

$$\frac{d^2u_1}{d\theta^2} + u_1 = \frac{3GM}{c^2}u_0^2$$

Because u_0 contains a $\cos\theta$ term, the source u_0^2 contains a component proportional to $\cos\theta$, which resonates with the homogeneous solution of the left-hand side and produces a slow rotation of the perihelion.

Solving the perturbed equation in the usual way, one finds that the radial oscillation acquires a slightly shifted angular frequency, so that after one full revolution the perihelion has advanced by an angle

$$\Delta\theta = \frac{6\pi GM}{a(1 - e^2)c^2}$$

where a is the semi-major axis of the orbit and e its eccentricity.

This result coincides exactly with the standard Einstein prediction for the relativistic perihelion advance, but here it arises solely from the effective radial force law induced by the orbital potential in flat Minkowski spacetime.

The appearance of the correct relativistic perihelion advance within a strictly flat-spacetime framework is particularly striking. No curvature, metric perturbation, or geodesic deviation has been invoked at any stage: the entire effect emerges from a single modification of the radial force law, generated purely by the orbital potential and its coupling to the conserved angular momentum. The post–Newtonian term $3GMh^2/(c^2r^4)$, derived algebraically in the previous section, is sufficient on its own to shift the frequency of the orbital oscillation and produce exactly the same precession predicted by general relativity. This illustrates in the clearest possible way the power of the effective-field approach: relativistic corrections traditionally attributed to curvature arise here from a linear, Maxwell-like structure in Minkowski spacetime, with the orbital potential providing the missing dynamical ingredient that links the source field to the geometry of the trajectory itself.

5.3 Light Deflection

5.3.1 Small–angle deflection formula

Consider a central potential $\phi(r)$ and a test body approaching from $x = -\infty$ with asymptotic proper–velocity magnitude $u = |\mathbf{u}|$ and impact parameter b. Along the unperturbed Minkowskian straight line we may write

$$r(x, b) = \sqrt{x^2 + b^2}, \qquad dx = u\,d\tau$$

where τ is the proper time of the test body. In the rest frame of the source (note that in the present weak–field regime the longitudinal component of $\mathbf{u}$ remains essentially constant, so that only the transverse component changes) the flat–spacetime gravitational dynamics gives

$$\frac{d\mathbf{u}_\perp}{d\tau} = \mathbf{a}_\perp = -\frac{\partial\phi}{\partial b}\,\mathbf{e}_b$$

with $\mathbf{e}_b$ the unit vector along the impact parameter. The total transverse impulse in proper–velocity is then

$$\Delta\mathbf{u}_\perp = \int_{-\infty}^{+\infty} \mathbf{a}_\perp\,d\tau = -\int_{-\infty}^{+\infty} \frac{\partial\phi}{\partial b}\,\mathbf{e}_b\,d\tau$$

Using $dx = u\,d\tau$ along the unperturbed straight line, we obtain

$$\Delta\mathbf{u}_\perp = -\frac{1}{u}\int_{-\infty}^{+\infty}\frac{\partial\phi}{\partial b}\,\mathbf{e}_b\,dx$$

For small deflections the scattering angle is given by the ratio of the transverse to the longitudinal proper–velocity,

$$\Delta\theta \simeq \frac{|\Delta\mathbf{u}_\perp|}{u}$$

so that

$$\Delta\theta = -\frac{1}{u^2}\int_{-\infty}^{+\infty}\frac{\partial\phi}{\partial b}\,dx = -\frac{2}{u^2}\int_{0}^{+\infty}\frac{\partial\phi}{\partial b}\,dx$$

This is the small–angle deflection formula expressed directly in terms of the proper–velocity u. In the non–relativistic regime $u \simeq v$ and it reduces to the familiar Newtonian expression.

This formula will now be applied separately to the source–generated potential and to the orbital potential $\phi^{(\mathrm{orb})}$.

5.3.2 Sourceful contribution

For the source potential

$$\phi(r) = -\frac{GM}{r}, \qquad r = \sqrt{x^2+b^2}$$

we find

$$\frac{\partial\phi}{\partial b} = \frac{GM\,b}{(x^2+b^2)^{3/2}}$$

The small–angle formula the previous equation gives

$$\Delta\theta_S = -\frac{2}{u^2}\int_{0}^{+\infty}\frac{GM\,b}{(x^2+b^2)^{3/2}}\,dx$$

The integral is standard:

$$\int_{0}^{+\infty}\frac{dx}{(x^2+b^2)^{3/2}} = \frac{1}{b^2}$$

so that

$$\Delta\theta_S = \frac{2GM}{b\,u^2}$$

Using $u = \gamma v$, this may also be written as

$$\Delta\theta_S = \frac{2GM}{b\,\gamma^2 v^2}$$

For slow particles ($\gamma \simeq 1,\ u \simeq v$) this reproduces the standard Newtonian result, while for ultra–relativistic motion the sourceful contribution is strongly suppressed as $1/u^2$.

5.3.3 Deflection from the orbital potential $\phi^{(\mathrm{orb})}$

The effective flat–spacetime gravitational dynamics introduces the scalar orbital potential

$$\phi^{(\mathrm{orb})}(r) = -\frac{GM\,h^2}{c^2 r^3}$$

where h is the conserved specific angular momentum,

$$\mathbf{h} = \mathbf{r} \times \mathbf{u}, \qquad h = |\mathbf{h}|$$

At fixed h we have

$$\frac{\partial \phi^{(\mathrm{orb})}}{\partial b} = -\frac{GM\,h^2}{c^2}\frac{\partial}{\partial b}(r^{-3}) = \frac{3GM\,h^2\,b}{c^2\,(x^2+b^2)^{5/2}}$$

Substituting into the previous equation,

$$\Delta\theta_{\mathrm{orb}} = -\frac{2}{u^2}\int_0^{+\infty} \frac{3GM\,h^2\,b}{c^2\,(x^2+b^2)^{5/2}}\,dx$$

The integral

$$\int_0^{+\infty} \frac{dx}{(x^2+b^2)^{5/2}} = \frac{2}{3b^4}$$

yields

$$\Delta\theta_{\text{orb}} = \frac{4GM\,h^2}{c^2\,u^2\,b^3}$$

For a nearly straight trajectory we have

$$h = b\,u \qquad \Rightarrow \qquad \frac{h^2}{u^2 b^3} = \frac{1}{b}$$

so that

$$\Delta\theta_{\text{orb}} = \frac{4GM}{bc^2}$$

This result is remarkable: the entire orbital contribution is independent of the magnitude of the proper–velocity u. It represents an intrinsic bending of the Minkowskian straight line produced by the effective $1/r^3$ interaction encoded in $\phi^{(\text{orb})}$.

5.3.4 Massive particles versus light

For any body with nonzero rest mass and finite proper–velocity u, both terms

$$\Delta\theta_S = \frac{2GM}{b\,u^2}, \qquad \Delta\theta_{\text{orb}} = \frac{4GM}{bc^2}$$

contribute to the deflection.

In the ultra–relativistic regime, however, the situation changes. Since

$$u = \gamma v, \qquad \gamma = \frac{1}{\sqrt{1 - v^2/c^2}}$$

we have $u \to \infty$ as $v \to c$. Consequently,

$$\Delta\theta_S = \frac{2GM}{b\,u^2} \to 0 \qquad \text{for} \quad v \to c$$

so that the sourceful contribution is automatically quenched in the null limit by the $1/u^2$ factor.

The orbital term, on the other hand,

$$\Delta\theta_{\text{orb}} = \frac{4GM}{bc^2}$$

remains finite and universal. It survives unchanged in the limit $v \to c$ and provides the entire deflection angle for light rays. Thus, in the Flat Spacetime Gravity framework the full Einstein light–deflection angle

$$\Delta\theta_{\text{light}} = \frac{4GM}{bc^2}$$

emerges purely from the orbital potential, while the sourceful part is dynamically suppressed by the relativistic growth of the proper–velocity.

5.3.5 Summary

The full relativistic bending of light thus arises in a completely flat Minkowskian framework. No curvature of spacetime, no geodesic equations, and no metric deformation are required. The effect originates algebraically from the orbital potential

$$\phi^{(\text{orb})}(r) = -\frac{GMh^2}{c^2r^3}$$

whose $1/c^2$ structure induces a genuine, velocity–independent distortion of straight Minkowskian trajectories near a massive body and it reproduces a phenomenon traditionally attributed to curved geometry. The agreement with the Einstein light–deflection angle therefore emerges as a direct consequence of the linear, flat-spacetime gravitational dynamics.

5.4 Gravitational Redshift

5.4.1 Static observers and effective energy

For static observers at fixed radius r we have

$$\mathbf{u} = \mathbf{0}, \qquad E = 0$$

so, the effective energy relation we have proved in the previous section

$$\gamma(r) = 1 - \frac{\phi^{(eff)}(r)}{c^2} + \frac{E}{c^2}$$

reduces to

$$\gamma(r) = 1 - \frac{\phi^{(\mathrm{eff})}(r)}{c^2}$$

The proper time τ of the static clock and the coordinate time t are related by

$$\frac{dt}{d\tau} = \gamma(r), \qquad \frac{d\tau}{dt} = \frac{1}{\gamma(r)}$$

In the weak–field regime $|\phi^{(\mathrm{eff})}|/c^2 \ll 1$ we expand

$$\begin{aligned} \gamma(r) &\simeq 1 - \frac{\phi^{(\mathrm{eff})}(r)}{c^2} \\ \Rightarrow \quad \frac{d\tau}{dt} &= \frac{1}{\gamma(r)} \simeq 1 + \frac{\phi^{(\mathrm{eff})}(r)}{c^2} \end{aligned}$$

Thus, a static clock at radius r ticks at the rate

$$\left.\frac{d\tau}{dt}\right|_r \simeq 1 + \frac{\phi^{(\mathrm{eff})}(r)}{c^2}$$

5.4.2 Relative ticking rates and frequency shift

Consider two static clocks at radii r_1 and r_2. To first order in $|\phi^{(\mathrm{eff})}|/c^2$,

$$\begin{aligned} \left.\frac{d\tau_1}{dt}\right|_{r_1} &\simeq 1 + \frac{\phi^{(\mathrm{eff})}(r_1)}{c^2} \\ \left.\frac{d\tau_2}{dt}\right|_{r_2} &\simeq 1 + \frac{\phi^{(\mathrm{eff})}(r_2)}{c^2} \end{aligned}$$

If a light signal of frequency ν_1 is emitted at r_1 and received at r_2, the number of wave crests is invariant, so the frequency measured by each observer is inversely proportional to their local proper-time rate. Therefore, the fractional frequency shift is

$$\frac{\Delta\nu}{\nu} = \frac{\nu_2 - \nu_1}{\nu_1} \simeq \frac{\phi^{(\text{eff})}(r_1) - \phi^{(\text{eff})}(r_2)}{c^2}$$

For static clocks the orbital term vanishes ($h = 0$), so that

$$\begin{aligned} \phi^{(\text{eff})}(r) &= \phi(r) + \phi^{(\text{orb})}(r) \\ &\Rightarrow \phi^{(\text{eff})}(r) = \phi(r) \end{aligned}$$

and in particular, for a central mass M,

$$\phi(r) = -\frac{GM}{r}$$

Hence the flat-spacetime gravitational theory reproduces the familiar weak–field gravitational redshift formula

$$\frac{\Delta\nu}{\nu} \simeq \frac{\phi(r_1) - \phi(r_2)}{c^2}$$

identical to the prediction of general relativity in the Schwarzschild weak–field limit.

5.4.3 Summary

The gravitational redshift thus emerges in the flat-spacetime gravitational framework without any reference to curved spacetime or metric deformation. It follows directly from the flat–space effective energy relation $\gamma c^2 + \phi^{(\text{eff})} = \text{const}$ and the kinematical link between proper time and coordinate time. For static observers the orbital contribution vanishes identically and the standard redshift depends solely on the Newtonian potential, showing that, at this level, the Flat Spacetime Gravity reproduces the classical general–relativistic result using nothing more than Minkowskian geometry and the effective potential formalism.

5.5 Shapiro Time Delay

5.5.1 Effective coordinate speed of light

To describe the Shapiro delay we need an effective propagation speed for light rays in the static field. For any worldline the relations

$$dt = \gamma\, d\tau, \qquad d\ell = \frac{1}{\gamma}\, d\ell_0$$

hold, where $d\ell_0$ is the proper spatial element and $d\ell$ the coordinate spatial element. For light in flat spacetime $d\ell_0/d\tau = c$, and therefore in the presence of the field the coordinate speed is

$$c_{\text{eff}} = \frac{d\ell}{dt} = \frac{d\ell_0/\gamma}{\gamma\, d\tau} = \frac{c}{\gamma^2}$$

As for the case of redshift, for static observers at fixed radius r we have

$$\mathbf{u} = \mathbf{0}, \qquad E = 0$$

so, the effective energy relation we have proved in the previous section

$$\gamma(r) = 1 - \frac{\phi^{(eff)}(r)}{c^2} + \frac{E}{c^2}$$

reduces to

$$\gamma(r) = 1 - \frac{\phi^{(\text{eff})}(r)}{c^2}$$

and thus

$$c_{\text{eff}}(r) \simeq \frac{c}{1 - 2\phi^{(\text{eff})}(r)/c^2}$$

For light travelling along a quasi–straight path in the static field, the coordinate travel time is

$$t = \int \frac{d\ell}{c_{\text{eff}}(r)} \simeq \frac{1}{c}\int d\ell \left(1 - \frac{2\phi^{(\text{eff})}(r)}{c^2}\right)$$

where we have expanded to first order in $|\phi^{(\mathrm{eff})}|/c^2$. The first term gives the flat–spacetime time of flight t_0; the second term gives the Shapiro delay

$$\Delta t = t - t_0 = -\frac{2}{c^3}\int \phi^{(\mathrm{eff})}(r)\,d\ell$$

5.5.2 Integration along the path

Along the near–straight path past a mass M with impact parameter b we may write

$$r^2 = x^2 + b^2, \qquad d\ell \simeq dx$$

where x is the coordinate along the unperturbed ray. The effective potential can be decomposed as

$$\phi^{(\mathrm{eff})}(r) = \phi(r) + \phi^{(\mathrm{orb})}(r)$$

For the Shapiro experiment the orbital term $\phi^{(\mathrm{orb})}$ contributes only a constant shift along the path (it does not depend on the detailed geometry of the near–grazing segment in a way that distinguishes different configurations), and therefore does not affect the *relative* delay between different trajectories. We may thus take

$$\phi^{(\mathrm{eff})}(r) \simeq \phi(r) \qquad \phi(r) = -\frac{GM}{r}$$

Then

$$\Delta t \simeq -\frac{2}{c^3}\int\left(-\frac{GM}{r}\right)d\ell = \frac{2GM}{c^3}\int_{x_1}^{x_2}\frac{dx}{\sqrt{x^2+b^2}}.$$

The integral is elementary:

$$\int \frac{dx}{\sqrt{x^2+b^2}} = \ln\left(x+\sqrt{x^2+b^2}\right) + \mathrm{const}$$

so that

$$\Delta t \simeq \frac{2GM}{c^3}\left[\ln(x + \sqrt{x^2 + b^2})\right]_{x_1}^{x_2}$$

Writing $r_1 = \sqrt{x_1^2 + b^2}$ and $r_2 = \sqrt{x_2^2 + b^2}$ and adopting the usual observational geometry with $r_1, r_2 \gg b$ and $x_1 \approx -r_1,\ x_2 \approx +r_2$, one finds to leading order

$$\Delta t \simeq \frac{4GM}{c^3}\ln\left(\frac{4r_1 r_2}{b^2}\right)$$

This is exactly the standard Shapiro time delay of the Schwarzschild solution in the weak–field limit.

5.5.3 Summary

In the flat-spacetime gravitational framework, the Shapiro delay arises solely from the effective coordinate speed $c_{\text{eff}} = c/\gamma^2$ implied by the flat-spacetime relation between dt, $d\tau$ and $d\ell$, together with the weak–field dependence $\gamma = 1 - \phi^{(\text{eff})}/c^2$. No curved metric, null geodesics or spacetime shearing are required: the logarithmic excess travel time is a direct consequence of Minkowskian geometry together with the effective potential itself.

5.6 Concluding Remarks

The four classical weak–field tests—perihelion advance, light deflection, gravitational redshift, and Shapiro time delay—constitute the most stringent observational benchmarks for any relativistic theory of gravitation. In the flat-spacetime gravitational framework each of these effects arises naturally and unavoidably from the same underlying ingredient: the effective potential

$$\phi^{(\text{eff})} = \phi + \phi^{(\text{orb})}$$

combined with the flat–spacetime kinematics and the dynamical force law. No curved metric, no perturbative expansion of a background geometry, and no geodesic deviation are required. Instead, the effective dynamical force law supplies precisely the additional $1/c^2$ terms that, in general relativity, are encoded geometrically within the Schwarzschild metric.

Despite this conceptual difference, the observational predictions coincide exactly with those of general relativity in the weak–field regime. The $3GMh^2/(c^2r^4)$ correction to the radial force fully accounts for the relativistic perihelion shift; the velocity–independent bending of Minkowskian trajectories generated by $\phi^{(\mathrm{orb})}$ yields the correct light–deflection angle; the relation between proper time and coordinate time reproduces the standard gravitational redshift; and the effective coordinate speed $c_{\mathrm{eff}} = c/\gamma^2$ produces the familiar logarithmic Shapiro delay. In every case the agreement with Einstein's theory emerges from the internal logic of special relativity combined with the flat-spacetime gravitational coupling, rather than from curvature of spacetime.

Thus, the astronomical tests do not merely confirm the consistency of the flat-spacetime gravitational theory with observation; they highlight a deeper point: key relativistic gravitational phenomena can be reproduced in a strictly Minkowskian setting once the orbital contribution is taken into account. The effective field encodes the required dynamical corrections through a linear, Maxwell–type structure, demonstrating that curvature is not the only route to capturing the observed behavior of gravity in the weak–field domain.

6. Relativistic Lagrangian Formulation

6.1 Introductory Remark

The effective flat-spacetime gravitational dynamics developed in previous sections can be placed on a fully covariant variational footing. In complete analogy with relativistic electrodynamics, the motion of a test particle follows from a worldline action composed of three contributions:

- the free relativistic particle action,
- the minimal coupling to the effective gravitational potential $\Phi_{\mu}^{(\text{eff})}$,
- the Maxwell–type field action associated with the antisymmetric gravitational field tensor $G_{\mu\nu}$.

This splitting parallels exactly the structure of the standard action of relativistic electrodynamics, with the substitution $qA_{\mu} \to -m_0\, \Phi_{\mu}^{(\text{eff})}$. The resulting variational principle is implemented in proper time τ, and the corresponding Euler–Lagrange equations are subsequently reduced to ordinary coordinate time t using $dt/d\tau = \gamma$. This reparametrization removes unwanted factors of γ from the particle Lagrangian and yields the usual d/dt Euler–Lagrange equations familiar from classical mechanics.

6.2 Total Action and Variational Principle

The total Lagrangian and the total action is the sum of the particle contribution and the field contribution:

$$L_{\text{tot}} = L_{\text{part}} + L_{\text{field}}, \qquad S_{\text{tot}} = S_{\text{part}} + S_{\text{field}}$$

$$L_{\text{part}} = L_{\text{free}} + L_{\text{int}}, \qquad S_{\text{part}} = S_{\text{free}} + S_{\text{int}}$$

Free particle

Because the parametrization here is in proper time, the free particle Lagrangian reduces to the constant $L_{\text{free}} = -m_0c^2$

$$L_{\text{free}} = -m_0c^2$$
$$S_{\text{free}} = -m_0c^2 \int d\tau$$

$$d\tau^2 = -\frac{1}{c^2}\eta_{\mu\nu}\, dx^\mu dx^\nu$$

Minimal coupling to the effective potential

$$L_{\text{int}} = m_0 \Phi_\mu^{(\text{eff})}\, U^\mu$$
$$S_{\text{int}} = m_0 \int \Phi_\mu^{(\text{eff})}\, U^\mu\, d\tau$$

$$U^\mu = \frac{dx^\mu}{d\tau}$$

Maxwell–type field

$$L_{\text{field}} = -\frac{c^2}{16\pi G} G_{\mu\nu}G^{\mu\nu}$$
$$S_{\text{field}} = -\frac{c^2}{16\pi G}\int G_{\mu\nu}G^{\mu\nu}\, d^4x$$

where

$$G_{\mu\nu} = \partial_\mu\Phi_\nu - \partial_\nu\Phi_\mu$$

Remark

The L_{int} interaction term $+\, m_0\, \Phi_\mu U^\mu$ carries the same overall structure as the electromagnetic term $+\, q\, A_\mu U^\mu$, although the dictionary says $q \mapsto -m_0$. This is because, as we have seen in the subsection of sign conventions in the charge–mass correspondence, the sign flip has already been taken into account in the gravitational potential (via $\phi < 0$ for attractive Newtonian fields). Therefore, no second sign flip should be applied at the level of the Lagrangian coupling.

Variational principle

The worldline of the particle is obtained by extremizing S_{part}:

$$\delta S_{\text{part}} = 0$$

under fixed endpoints $\delta x^{\mu}(\tau_1) = \delta x^{\mu}(\tau_2) = 0$. This yields the covariant Euler–Lagrange equations

$$\frac{d}{d\tau}\left(\frac{\partial L_{\text{part}}}{\partial U^{\mu}}\right) - \frac{\partial L_{\text{part}}}{\partial x^{\mu}} = 0$$

6.3 Reparameterization: from $d\tau$ to dt

To obtain evolution equations in ordinary coordinate time, we perform the reparameterization

$$d\tau = \frac{dt}{\gamma}, \qquad \gamma = \frac{dt}{d\tau}$$

and write

$$S_{\text{part}} = \int L_{\text{part}}\, d\tau = \int \frac{L_{\text{part}}}{\gamma}\, dt = \int \mathcal{L}\, dt$$

Because the velocity-dependent part of L_{part} is homogeneous of degree one in U^{μ}, the factor $1/\gamma$ removes all explicit γ – dependence from the parametrized particle Lagrangian. The resulting coordinate–time Lagrangian is

$$L_{\text{part}} = -m_0 c^2 + m_0\, \Phi_{\mu}^{(\text{eff})}\, U^{\mu} = -m_0 c^2 - m_0 \gamma\, \phi^{(\text{eff})} + m_0\, \mathbf{A}^{(\text{eff})} \cdot \mathbf{u}$$

$$\mathcal{L} = \frac{L_{\text{part}}}{\gamma} = -m_0 c^2 \sqrt{1 - \frac{v^2}{c^2}} - m_0\, \phi^{(\text{eff})} + m_0\, \mathbf{A}^{(\text{eff})} \cdot \mathbf{v}$$

where

$$\Phi_{\mu}^{(\text{eff})} = \left(-\frac{\phi^{(\text{eff})}}{c},\ \mathbf{A}^{(\text{eff})}\right)$$

The Euler–Lagrange equations now take their standard t–parametrized form:

$$\frac{d}{dt}\left(\frac{\partial \mathcal{L}}{\partial \mathbf{v}}\right) - \frac{\partial \mathcal{L}}{\partial \mathbf{x}} = 0$$

and will reproduce exactly the effective gravitational force law derived in previous section.

6.4 Covariant derivation of the effective dynamical law

6.4.1 Variation of the particle action

We now show that the axiomatic dynamical law

$$a^{\mu} = \mathrm{G}_{\nu}^{(\text{eff})\mu}\, U^{\nu}$$

follows directly from the relativistic particle action in the effective potential $\Phi_{\mu}^{(\text{eff})}$.

As we have seen, the particle action is

$$S_{\text{part}} = -m_0 c^2 \int \mathrm{d}\tau + m_0 \int \Phi_{\mu}^{(\text{eff})}\, U^{\mu}\, \mathrm{d}\tau, \quad U^{\mu} \equiv \frac{\mathrm{d}x^{\mu}}{\mathrm{d}\tau}$$

We vary the worldline $x^{\mu}(\tau) \rightarrow x^{\mu}(\tau) + \delta x^{\mu}(\tau)$, keeping the endpoints fixed: $\delta x^{\mu}(\tau_1) = \delta x^{\mu}(\tau_2) = 0$. The variation of the four–velocity is

$$\delta U^{\mu} = \frac{\mathrm{d}}{\mathrm{d}\tau}\, \delta x^{\mu}$$

Free part

The variation of the free term $-m_0 c^2 \int \mathrm{d}\tau$ is standard: one finds

$$\delta S_{\text{free}} = -m_0 \int a_{\mu}\, \delta x^{\mu}\, \mathrm{d}\tau, \qquad a^{\mu} \equiv \frac{\mathrm{d}U^{\mu}}{\mathrm{d}\tau}$$

so, the free particle contributes the usual inertial term $m_0 a^{\mu}$ to the equations of motion.

Interaction part

For the interaction term

$$S_{\text{int}} = m_0 \int \Phi_{\mu}^{(\text{eff})}(x)\, U^{\mu}\, \mathrm{d}\tau$$

we obtain

$$\delta S_{\text{int}} = m_0 \int \left[\left(\partial_{\nu}\, \Phi_{\mu}^{(\text{eff})} \right) \delta x^{\nu}\, U^{\mu} + \Phi_{\mu}^{(\text{eff})}\, \delta U^{\mu} \right] \mathrm{d}\tau$$

Using $\delta U^{\mu} = \mathrm{d}(\delta x^{\mu})/\mathrm{d}\tau$ and integrating by parts,

$$\int \Phi_{\mu}^{(\text{eff})}\, \delta U^{\mu}\, \mathrm{d}\tau = [\Phi_{\mu}^{(\text{eff})}\, \delta x^{\mu}]_{\tau_1}^{\tau_2} - \int \frac{\mathrm{d}\Phi^{(\text{eff})}{}_{\mu}}{\mathrm{d}\tau}\, \delta x^{\mu}\, \mathrm{d}\tau$$

The boundary term vanishes (fixed endpoints), so

$$\delta S_{\text{int}} = m_0 \int \left[(\partial_{\nu}\, \Phi_{\mu}^{(\text{eff})})\, U^{\mu} - \frac{\mathrm{d}\, \Phi_{\nu}^{(\text{eff})}}{\mathrm{d}\tau} \right] \delta x^{\nu}\, \mathrm{d}\tau$$

But

$$\frac{\mathrm{d}\, \Phi_{\nu}^{(\text{eff})}}{\mathrm{d}\tau} = \partial_{\mu}\, \Phi_{\nu}^{(\text{eff})}\, U^{\mu}$$

so, the bracket simplifies to

$$\left(\partial_\nu \Phi^{(\text{eff})}_\mu\right) U^\mu - \left(\partial_\mu \Phi^{(\text{eff})}_\nu\right) U^\mu$$
$$= \left(\partial_\nu \Phi^{(\text{eff})}_\mu - \partial_\mu \Phi^{(\text{eff})}_\nu\right) U^\mu = G^{(\text{eff})}_{\nu\mu} U^\mu$$

Hence

$$\delta S_{\text{int}} = m_0 \int G^{(\text{eff})}_{\nu\mu} U^\mu \, \delta x^\nu \, \mathrm{d}\tau$$

6.4.2 Euler–Lagrange equation and effective force law

The total variation of the particle action is

$$\delta S_{\text{part}} = -m_0 \int \left[a_\nu - G^{(\text{eff})}_{\nu\mu} U^\mu\right] \delta x^\nu \, \mathrm{d}\tau$$

Requiring $\delta S_{\text{part}} = 0$ for arbitrary $\delta x^\nu(\tau)$ gives

$$a_\nu = G^{(\text{eff})}_{\nu\mu} U^\mu$$

Raising the index with the Minkowski metric $\eta^{\mu\nu}$,

$$a^\mu = \mathrm{G}^{(\text{eff})\mu}_{\nu} U^\nu$$

which is precisely the covariant dynamical law postulated in previous sections.

6.4.3 Orthogonality and preservation of the mass shell

Because $G^{(\text{eff})}_{\mu\nu}$ is antisymmetric,

$$U_\mu a^\mu = U_\mu \mathrm{G}^{(\text{eff})\mu}_{\nu} U^\nu = \frac{1}{2} G^{(\text{eff})}_{\mu\nu} U^\mu U^\nu = 0$$

so

$$\frac{\mathrm{d}}{\mathrm{d}\tau}\left(U_\mu U^\mu\right) = 2\, U_\mu a^\mu = 0$$

Thus, the normalization condition $U_\mu U^\mu = -c^2$ is preserved along the worldline: the effective field $G_{\mu\nu}^{(\text{eff})}$ changes only the *direction* of U^μ, never its invariant magnitude. This completes the fully covariant derivation of the dynamical law

$$a^\mu = G_\nu^{(\text{eff})\mu} U^\nu$$

from the relativistic particle Lagrangian in flat spacetime.

6.5 Energy conservation

Because the coordinate–time Lagrangian

$$\mathcal{L} = -m_0 c^2 \sqrt{1 - \frac{v^2}{c^2}} - m_0\, \phi^{(\text{eff})} + m_0\, \mathbf{A}^{(\text{eff})} \cdot \mathbf{v}$$

contains no explicit dependence on t, Noether's theorem guarantees that the energy

$$H \equiv \mathbf{v} \cdot \frac{\partial \mathcal{L}}{\partial \mathbf{v}} - \mathcal{L}$$

is a conserved quantity.

We compute

$$\frac{\partial \mathcal{L}}{\partial \mathbf{v}} = m_0 \gamma\, \mathbf{v} + m_0\, \mathbf{A}^{(\text{eff})}$$

so that

$$\mathbf{v} \cdot \frac{\partial \mathcal{L}}{\partial \mathbf{v}} = m_0 \gamma\, v^2 + m_0\, \mathbf{A}^{(\text{eff})} \cdot \mathbf{v}$$

Inserting this expression into Noether's definition gives

$$H = m_0 \gamma v^2 + m_0\, \mathbf{A}^{(\text{eff})} \cdot \mathbf{v}$$
$$- \left[-m_0 c^2 \sqrt{1 - \frac{v^2}{c^2}} - m_0\, \phi^{(\text{eff})} + m_0\, \mathbf{A}^{(\text{eff})} \cdot \mathbf{v} \right]$$

The $\mathbf{A}^{(\text{eff})} \cdot \mathbf{v}$ terms cancel, and we obtain

$$H \;=\; m_0 \gamma v^2 + m_0 c^2 \sqrt{1 - \frac{v^2}{c^2}} + m_0\, \phi^{(\text{eff})}$$

Using the identities

$$\gamma^2 v^2 = c^2(\gamma^2 - 1), \qquad \gamma^{-1} = \sqrt{1 - \frac{v^2}{c^2}}$$

the expression simplifies dramatically:

$$H \;=\; m_0 \gamma c^2 + m_0 \phi^{(\text{eff})} \;=\; \text{constant}$$

Setting the constant H to be

$$H \;=\; m_0\, c^2 + m_0 E$$

we obtain

$$m_0 E \;=\; m c^2 - m_0 c^2 + m_0 \phi^{(\text{eff})}, \qquad m = \; m_0 \gamma$$

$$E = \gamma c^2 - c^2 + \phi^{(\text{eff})}$$

and solving for γ gives the linear relation

$$\gamma \;=\; 1 - \frac{\phi^{(\text{eff})}}{c^2} + \frac{E}{c^2}$$

which matches exactly the effective energy–potential relation previously obtained from the dynamical law. Noether's theorem therefore provides a fully independent variational derivation of the same conservation principle.

6.6 Conservation of generalized momentum

6.6.1 Reparameterization

To derive the conservation of generalized momentum it is convenient to start from the reparameterization–invariant form of the particle action,

$$S_{\text{part}} = \int L\, d\lambda$$

$$L = -m_0 c \sqrt{-U_\mu U^\mu} + m_0\, \Phi_\mu^{(\text{eff})} U^\mu$$

where

$$U^\mu \equiv \frac{dx^\mu}{d\lambda}, \qquad U_\mu U^\mu < 0$$

Proper time τ is obtained by choosing the gauge

$$\sqrt{-U_\mu U^\mu} = c$$

in which case U^μ coincides with the usual four–velocity.

6.6.2 Noether current for spacetime translations

Consider an infinitesimal spacetime translation along the coordinate x^α,

$$x^\mu \to x^\mu + \varepsilon\, \delta^\mu_\alpha, \qquad \delta x^\mu = \varepsilon\, \delta^\mu_\alpha, \quad \delta U^\mu = \frac{d}{d\lambda}(\delta x^\mu) = 0$$

The Lagrangian changes as

$$\delta L = \frac{\partial L}{\partial x^\mu}\, \delta x^\mu = \varepsilon\, \frac{\partial L}{\partial x^\alpha}$$

If the effective potential is *homogeneous* in x^α,

$$\partial_\alpha \Phi_\mu^{(\text{eff})} = 0$$

then L has no explicit dependence on x^α and $\partial L/\partial x^\alpha = 0$. The action is invariant under this translation, and Noether's theorem guarantees the existence of a conserved quantity.

The Noether charge associated with the transformation $\delta x^\mu = \varepsilon\, \delta^\mu_\alpha$ is

$$Q_\alpha = \frac{\partial L}{\partial U^\mu}\, \delta x^\mu = \varepsilon\, \frac{\partial L}{\partial U^\alpha}$$

Thus, the relevant quantity is

$$\frac{\partial L}{\partial U^\alpha} = -m_0 c \frac{\partial}{\partial U^\alpha}\sqrt{-U_\mu U^\mu} + m_0\, \Phi_\alpha^{(\mathrm{eff})}$$

Using

$$\sqrt{-U_\mu U^\mu} = \sqrt{-U^2}, \qquad \frac{\partial}{\partial U^\alpha}\sqrt{-U^2} = -\frac{U_\alpha}{\sqrt{-U^2}}$$

we obtain

$$\frac{\partial L}{\partial U^\alpha} = m_0 c \frac{U_\alpha}{\sqrt{-U^2}} + m_0\, \Phi_\alpha^{(\mathrm{eff})}$$

In the proper–time gauge $\sqrt{-U^2} = c$ this simplifies to

$$\frac{\partial L}{\partial U^\alpha} = m_0 \left(U_\alpha + \Phi_\alpha^{(\mathrm{eff})}\right)$$

Noether's theorem then gives

$$\frac{d}{d\lambda}\left(\frac{\partial L}{\partial U^\alpha}\right) = 0 \quad \Rightarrow \quad \frac{d}{d\lambda}\left(U_\alpha + \Phi_\alpha^{(\mathrm{eff})}\right) = 0$$

Choosing $\lambda = \tau$, we finally obtain the covariant conservation law

$$\frac{d}{d\tau}\left(U_\alpha + \Phi_\alpha^{(\mathrm{eff})}\right) = 0 \qquad (\text{if } \partial_\alpha \Phi_\mu^{(\mathrm{eff})} = 0)$$

6.6.3 Interpretation as generalized momentum

The quantity

$$P_\alpha^{(\mathrm{eff})} = m_0 \left(U_\alpha + \Phi_\alpha^{(\mathrm{eff})}\right)$$

plays the role of a *generalized momentum* conjugate to x^α: it is conserved precisely when the effective potential is homogeneous in that coordinate. For $\alpha = 0$ (time translations) the conservation of $P_0^{(\mathrm{eff})}$ reproduces the energy conservation law we have seen earlier; for spatial directions, homogeneity in x^i implies the existence of conserved spatial generalized momenta associated with linear or angular invariances.

In this way Noether's theorem unifies the energy integral and the conserved momenta into a single covariant statement: whenever $\Phi_{\mu}^{(\mathrm{eff})}$ is independent of a coordinate x^{α}, the corresponding generalized momentum $P_{\alpha}^{(\mathrm{eff})}$ remains constant along the worldline.

6.7 Angular momentum and rotational symmetry

6.7.1 Infinitesimal rotations and Noether charge

The same relativistic Lagrangian

$$\mathcal{L}(t,\mathbf{x},\mathbf{v}) = -m_0c^2\sqrt{1-\frac{v^2}{c^2}} - m_0\,\phi^{(\mathrm{eff})}(t,\mathbf{x}) + m_0\,\mathbf{A}^{(\mathrm{eff})}(t,\mathbf{x})\cdot\mathbf{v}$$

used in the previous subsections, is invariant under spatial rotations whenever $\phi^{(\mathrm{eff})}$ and $\mathbf{A}^{(\mathrm{eff})}$ depend only on rotationally invariant combinations (e.g. the radius $r = |\mathbf{x}|$) and transform as a scalar field and a vector field, respectively. In that case Noether's theorem implies conservation of a generalized angular momentum.

An infinitesimal spatial rotation can be written as

$$\delta x_i = \omega_k\,\varepsilon_{kij}\,x_j, \qquad \delta v_i = \omega_k\,\varepsilon_{kij}\,v_j, \qquad i,j,k = 1,2,3$$

where $\boldsymbol{\omega} = (\omega_1,\omega_2,\omega_3)$ is a constant infinitesimal parameter and ε_{kij} the Levi–Civita symbol. In vector notation

$$\delta\mathbf{x} = \boldsymbol{\omega}\times\mathbf{x}, \qquad \delta\mathbf{v} = \boldsymbol{\omega}\times\mathbf{v}$$

If $\phi^{(\mathrm{eff})}$ and $\mathbf{A}^{(\mathrm{eff})}$ transform as scalar and vector under rotations (and depend only on r), the Lagrangian is invariant:

$$\delta\mathcal{L} = \frac{\partial\mathcal{L}}{\partial\mathbf{x}}\cdot\delta\mathbf{x} + \frac{\partial\mathcal{L}}{\partial\mathbf{v}}\cdot\delta\mathbf{v} = 0$$

Noether's theorem then associates with this symmetry the conserved charge

$$Q[\boldsymbol{\omega}] = \frac{\partial \mathcal{L}}{\partial \mathbf{v}} \cdot \delta \mathbf{x} = \mathbf{p} \cdot (\boldsymbol{\omega} \times \mathbf{x})$$

where

$$\mathbf{p} \equiv \frac{\partial \mathcal{L}}{\partial \mathbf{v}} = m_0 \gamma \, \mathbf{v} + m_0 \mathbf{A}^{(\text{eff})}$$

is the canonical momentum. Using $\mathbf{p} \cdot (\boldsymbol{\omega} \times \mathbf{x}) = (\mathbf{x} \times \mathbf{p}) \cdot \boldsymbol{\omega}$, we obtain

$$Q[\boldsymbol{\omega}] = (\mathbf{x} \times \mathbf{p}) \cdot \boldsymbol{\omega}$$

Since $\boldsymbol{\omega}$ is arbitrary and constant, the conserved Noether quantity is the vector

$$\mathbf{l} = \mathbf{x} \times \mathbf{p}, \qquad \frac{d\mathbf{l}}{dt} = \mathbf{0}$$

Thus, rotational invariance implies conservation of the generalized angular momentum vector $\mathbf{l}$.

6.7.2 Relation to specific orbital angular momentum

In the flat-spacetime gravitational setting we have

$$\mathbf{p} = m_0 \gamma \, \mathbf{v} + m_0 \mathbf{A}^{(\text{eff})}$$

so that

$$\begin{aligned} \mathbf{l} &= \mathbf{x} \times (m_0 \gamma \, \mathbf{v}) + \mathbf{x} \times m_0 \mathbf{A}^{(\text{eff})} \\ &= m_0 \, (\mathbf{x} \times \mathbf{u}) + m_0 \mathbf{x} \times \mathbf{A}^{(\text{eff})}, \qquad \mathbf{u} = \gamma \mathbf{v} \end{aligned}$$

In the simplest case of a static, spherically symmetric configuration in the rest frame of the source, the effective vector potential vanishes, $\mathbf{A}^{(\text{eff})} = \mathbf{0}$, and we obtain

$$\mathbf{l} = m_0 \, \mathbf{h}, \qquad \mathbf{h} \equiv \mathbf{x} \times \mathbf{u}$$

The vector **h** is the specific orbital angular momentum. Since **l** is conserved, it follows that

$$\frac{d\mathbf{h}}{d\tau} = \mathbf{0}, \qquad h = |\mathbf{h}| = \text{const}$$

In planar motion one usually works with the scalar

$$h = r^2 \gamma \dot{\theta}$$

which is precisely the quantity entering the orbital potential $\phi^{(\mathrm{orb})}$ and the analysis of the perihelion advance. Rotational symmetry therefore guarantees the constancy of h along the orbit.

Thus, conservation of angular momentum in the flat-spacetime gravitational theory is encoded both in the Noether charge associated with spatial rotations,

$$\mathbf{l} = \mathbf{x} \times \mathbf{p}$$

and in the covariant bivector $\mathrm{H}_{\mu\nu}$. The spatial dual of $\mathrm{H}_{\mu\nu}$ is precisely the vector **h**, which plays a central role in the orbital sector of the theory and in the construction of the orbital potential $\phi^{(\mathrm{orb})}$.

6.8 Maxwell–type field equations from the variational principle

6.8.1 Field action

The dynamics of the fundamental flat-spacetime gravitational field is encoded in the antisymmetric tensor

$$G_{\mu\nu} = \partial_\mu \Phi_\nu - \partial_\nu \Phi_\mu, \qquad \Phi_\mu = \left(-\frac{\phi}{c}, \mathbf{A}\right)$$

defined on flat Minkowski spacetime. This Φ_μ is the *bare* gravitational potential generated by the mass sources.

On the other hand, the motion of a test particle is governed by the *effective* potential

$$\Phi_{\mu}^{(\mathrm{eff})} = \Phi_{\mu} + \Phi_{\mu}^{(\mathrm{orb})}$$

where $\Phi_{\mu}^{(\mathrm{orb})}$ is the orbital contribution constructed as in previous sections. It is a functional of the worldline and of $G_{\mu\nu}$, not an independent field degree of freedom. In the variational principle for the field, we therefore keep

$$\delta\Phi_{\mu}^{(\mathrm{orb})} \equiv 0, \qquad \delta\Phi_{\mu}^{(\mathrm{eff})} = \delta\Phi_{\mu}$$

The pure–field action is

$$S_{\mathrm{field}} = -\frac{c^2}{16\pi G}\int G_{\mu\nu}G^{\mu\nu}\, d^4x$$

with $G_{\mu\nu}$ built solely from Φ_{μ} as above.

The coupling to matter is written in terms of the *effective* potential:

$$S_{\mathrm{int}} = \int J^{\mu}\, \Phi_{\mu}^{(\mathrm{eff})}\, d^4x, \qquad J^{\mu} = \rho_0 U^{\mu}$$

where ρ_0 is the rest–mass density and U^{μ} the four–velocity field of the matter distribution.

The full field action is

$$S_F = S_{\mathrm{field}} + S_{\mathrm{int}} = \int d^4x \left[-\frac{c^2}{16\pi G}G_{\mu\nu}G^{\mu\nu} + J^{\mu}\Phi_{\mu}^{(\mathrm{eff})}\right]$$

6.8.2 Variation with respect to the bare potential Φ_{μ}

Since $\Phi_{\mu}^{(\mathrm{orb})}$ is not an independent field, we vary S_F with respect to the bare field Φ_{μ}, keeping $\Phi_{\mu}^{(\mathrm{orb})}$ fixed:

$$\delta\Phi_{\mu}^{(\mathrm{eff})} = \delta\Phi_{\mu}$$

The variation of the field term is

$$\delta G_{\mu\nu} = \partial_{\mu}\delta\Phi_{\nu} - \partial_{\nu}\delta\Phi_{\mu}$$

$$\delta S_{\text{field}} = -\frac{c^2}{16\pi G}\int 2\, G^{\mu\nu}\, \delta G_{\mu\nu}\, d^4x$$

$$= -\frac{c^2}{8\pi G}\int G^{\mu\nu}(\partial_\mu \delta\Phi_\nu - \partial_\nu \delta\Phi_\mu)\, d^4x$$

Integrating by parts and discarding boundary terms, and using the antisymmetry of $G^{\mu\nu}$,

$$\delta S_{\text{field}} = \frac{c^2}{4\pi G}\int (\partial_\mu G^{\mu\nu})\, \delta\Phi_\nu\, d^4x$$

For the interaction term we have

$$\delta S_{\text{int}} = \int J^\nu\, \delta\Phi_\nu^{(\text{eff})}\, d^4x = \int J^\nu\, \delta\Phi_\nu\, d^4x$$

since $\delta\Phi_\nu^{(\text{orb})} = 0$

6.8.3 Field equations

Non-Homogeneous equations

The full variation is

$$\delta S_F = \delta S_{\text{field}} + \delta S_{\text{int}} = \int \left[\frac{c^2}{4\pi G}\, \partial_\mu G^{\mu\nu} + J^\nu\right] \delta\Phi_\nu\, d^4x$$

Requiring $\delta S_F = 0$ for arbitrary $\delta\Phi_\nu$ yields

$$\frac{c^2}{4\pi G}\, \partial_\mu G^{\mu\nu} + J^\nu = 0$$

So

$$\partial_\mu G^{\mu\nu} = -\frac{4\pi G}{c^2}\, J^\nu$$

These are the inhomogeneous Maxwell–type equations for the *bare* field $G_{\mu\nu}$; the orbital contribution affects only the effective force law on test particles, not the source structure of the field itself.

Homogeneous equations

From the definition of $G_{\mu\nu}$ we obtain the Bianchi identity

$$\partial_\lambda G_{\mu\nu} + \partial_\mu G_{\nu\lambda} + \partial_\nu G_{\lambda\mu} = 0$$

which is the gravitational analogue of $\partial_{[\lambda} F_{\mu\nu]} = 0$ in electromagnetism

6.8.4 Wave equations for the potentials

Imposing the Lorenz gauge $\partial_\mu \Phi^\mu = 0$ and using $G_{\mu\nu} = \partial_\mu \Phi_\nu - \partial_\nu \Phi_\mu$, the inhomogeneous equation becomes

$$\Box \Phi^\nu = -\frac{4\pi G}{c^2} J^\nu$$

Thus, each component of Φ^ν satisfies a flat-spacetime wave equation.

6.8.5 Decomposition

Define the 3–vector fields

$$g_i = -c\, G_{0i}, \qquad G_{ij} = \varepsilon_{ijk} b_k$$

Decompose the source current as

$$J^\mu = (\rho_0 \gamma c, \mathbf{j}), \qquad \mathbf{j} = \rho_0 \gamma \mathbf{v}$$

Then the non-homogenous part gives

$$\begin{aligned} \nabla \cdot \mathbf{g} &= -4\pi G\, \rho_0 \gamma \\ \nabla \times \mathbf{b} &= -\frac{4\pi G}{c^2} \mathbf{j} + \frac{1}{c^2} \frac{\partial \mathbf{g}}{\partial t} \end{aligned}$$

while the homogeneous equations yield

$$\begin{aligned} \nabla \cdot \mathbf{b} &= 0 \\ \nabla \times \mathbf{g} &= -\frac{\partial \mathbf{b}}{\partial t} \end{aligned}$$

Together, these four equations form the complete Maxwell–type system for the flat-spacetime gravitational field $(\mathbf{g}, \mathbf{b})$. They are structurally identical to the Maxwell equations of electromagnetism with the replacements

$$(\mathbf{E}, \mathbf{B}) \to (\mathbf{g}, \mathbf{b}), \qquad \rho_{0,\mathrm{EM}} \to -\rho_0$$

The orbital potential $\Phi^{(\mathrm{orb})}{}_{\mu}$ modifies only the *effective* force law on test particles via

$$a^{\mu} = \mathrm{G}_{\nu}^{(\mathrm{eff})\mu}\, U^{\nu}$$

but does not change the source structure or the linearity of the field equations satisfied by Φ_{μ} and $G_{\mu\nu}$.

6.9 Field stress–energy tensor and its divergence

6.9.1 Field stress-energy tensor

In complete analogy with electromagnetism, the energy–momentum content of the flat-spacetime gravitational field is encoded in a symmetric rank–2 tensor constructed quadratically from the bare field tensor $G_{\mu\nu}$.

The field action is

$$S_{\mathrm{field}} = -\frac{c^2}{16\pi G}\int G_{\mu\nu}G^{\mu\nu}\, d^4x$$

Varying this action with respect to the Minkowski metric $\eta_{\mu\nu} = \mathrm{diag}(-1,1,1,1)$ yields the field stress–energy tensor

$$T^{\mu\nu}_{field} = \frac{c^2}{4\pi G}\left(G^{\mu\alpha}G^{\nu}_{\alpha} - \frac{1}{4}\eta^{\mu\nu}G^{\alpha\beta}G_{\alpha\beta}\right)$$

This is the direct flat-spacetime gravitational analogue of the electromagnetic stress–energy tensor, with the replacements

$$F_{\mu\nu} \to G_{\mu\nu}, \qquad \mu_0 \to \frac{4\pi G}{c^2}$$

Observe that the second term of the stress-energy tensor is simply

$$\frac{1}{4}\eta^{\mu\nu}G^{\alpha\beta}G_{\alpha\beta} = \frac{1}{4}\eta^{\mu\nu}\,\mathrm{I}_1$$

where is the first invariant I_1 of the flat-spacetime gravitational field tensor $G_{\mu\nu}$,

$$\mathrm{I}_1 \equiv \frac{1}{2}G_{\mu\nu}G^{\mu\nu} = |\mathbf{b}|^2 - \frac{|\mathbf{g}|^2}{c^2}$$

6.9.2 Energy density and flux

In a 3 + 1 split we use the standard identification

$$G_{0i} = -\frac{g_i}{c}, \qquad G_{ij} = \varepsilon_{ijk}\, b_k$$

where $\mathbf{g}$ is the sourceful gravitational field and $\mathbf{b}$ the rotational gravitational field. Inserting these into the expression of stress–energy tensor one obtains

$$T^{00}_{\text{field}} = \frac{c^2}{8\pi G}\left(\frac{|\mathbf{g}|^2}{c^2} + |\mathbf{b}|^2\right) = u_{\text{field}}$$

which matches the field energy density introduced in the purely three–dimensional formulation.

The mixed components describe the energy flux:

$$T^{0i}_{\text{field}} = \frac{c}{4\pi G}(\mathbf{g}\times\mathbf{b})_i = \frac{1}{c}\, S_i$$

with the flat-spacetime gravitational Poynting vector

$$\mathbf{S} = \frac{c^2}{4\pi G}\,\mathbf{g}\times\mathbf{b}$$

Thus u_{field} and $\mathbf{S}$ obey exactly the same structural relations as in electromagnetism, up to the replacement of the coupling constant.

6.9.3 Divergence of the field stress–energy tensor

Taking the divergence of the stress–energy tensor

$$\partial_\mu T^{\mu\nu}_{\text{field}} = \frac{c^2}{4\pi G}\,\partial_\mu\left(G^{\mu\alpha}G^{\nu}_{\alpha} - \frac{1}{4}\,\eta^{\mu\nu}G^{\alpha\beta}G_{\alpha\beta}\right)$$

we expand

$$\partial_\mu T^{\mu\nu}_{\text{field}} = \frac{c^2}{4\pi G}\left[(\partial_\mu G^{\mu\alpha})\,G^{\nu}_{\alpha} + G^{\mu\alpha}\,\partial_\mu G^{\nu}_{\alpha} - \frac{1}{4}\,\partial^\nu(G^{\alpha\beta}G_{\alpha\beta})\right]$$

The inhomogeneous Maxwell–type equations for the bare field,

$$\partial_\mu G^{\mu\alpha} = -\frac{4\pi G}{c^2}\,J^\alpha$$

allow us to rewrite the first term as

$$\frac{c^2}{4\pi G}\,(\partial_\mu G^{\mu\alpha})\,G^{\nu}_{\alpha} = -\,G^{\nu}_{\alpha}J^\alpha$$

The remaining two terms combine to zero by virtue of the antisymmetry of $G_{\mu\nu}$ (the standard identity is already familiar from relativistic electrodynamics),

$$G^{\mu\alpha}\,\partial_\mu G^{\nu}_{\alpha} - \frac{1}{4}\,\partial^\nu(G^{\alpha\beta}G_{\alpha\beta}) = 0$$

Hence the divergence reduces to

$$\partial_\mu T^{\mu\nu}_{\text{field}} = -\,G^{\nu}_{\alpha}\,J^\alpha$$

This is the flat-spacetime gravitational four–force density exerted by the field on the matter distribution.

6.9.4 Total stress–energy and local conservation

Let $T^{\mu\nu}_{\text{matter}}$ denote the stress–energy tensor of matter (treated here phenomenologically). The total stress–energy tensor is

$$T^{\mu\nu} = T^{\mu\nu}_{\text{matter}} + T^{\mu\nu}_{\text{field}}$$

Local conservation of total energy–momentum in flat spacetime is expressed by

$$\partial_\mu T^{\mu\nu} = 0$$

Thus

$$\partial_\mu T^{\mu\nu}_{\text{matter}} = +G^{\nu}_{\alpha}\, J^{\alpha}$$

which states that the change in matter energy–momentum density is driven by the flat-spacetime gravitational four–force density $G^{\nu}_{\alpha}\, J^{\alpha}$. The spatial components ($\nu = i$) yield relativistic Euler–type equations for the matter momentum density, while the temporal component ($\nu = 0$) encodes the local energy balance between matter and the flat-spacetime gravitational field.

Together with the single–particle force law

$$a^{\mu} = \mathrm{G}^{(\mathrm{eff})\mu}_{\nu} U^{\nu}$$

the above relations complete the field–theoretic description of energy and momentum exchange in the Flat Spacetime Gravity, in direct analogy with the electromagnetic case but entirely within a flat Minkowskian spacetime.

6.10 Concluding Remarks

The analysis carried out in this section establishes that flat-spacetime gravitational dynamics admits a complete and self–consistent Lagrangian formulation, fully parallel to that of classical electromagnetism. The bare gravitational potential Φ_μ enters the field action quadratically through the antisymmetric tensor $G_{\mu\nu}$, while matter couples linearly through the covariant current J^{μ}. Variation with respect to Φ_μ yields a Maxwell–type system of linear field equations, whose homogeneous and inhomogeneous parts follow uniquely from the definition of $G_{\mu\nu}$ and the principle of stationary action. At the same time, variation of the particle action produces a covariant force law of the Lorentz type,

$$a^{\mu} = \mathrm{G}_{\nu}^{(\mathrm{eff})\mu} U^{\nu}$$

demonstrating the full dynamical consistency between matter, sources, and fields.

Several structural features deserve emphasis. First, the theory retains the defining simplicity of a flat Minkowskian background: all quantities are tensors on $\eta_{\mu\nu}$, and no curved geometric structures are required in order to describe gravitational forces, radiation, or energy–momentum transfer. Second, the decomposition of the effective potential into a bare part and an orbital part,

$$\Phi_{\mu}^{(\mathrm{eff})} = \Phi_{\mu} + \Phi_{\mu}^{(\mathrm{orb})}$$

provides a natural and entirely algebraic mechanism by which the geometry of a worldline couples back into its own dynamics, without altering the linearity of the underlying field equations. In this respect the flat-spacetime gravitational framework mirrors the structure of electromagnetism more closely than any geometric theory of gravity, while still reproducing the physically relevant nonlinearities at the level of particle dynamics.

Finally, the stress–energy tensor

$$T_{\text{field}}^{\mu\nu} = \frac{c^2}{4\pi G}\left(G^{\mu\alpha} G_{\alpha}^{\nu} - \frac{1}{4}\eta^{\mu\nu} G^{\alpha\beta} G_{\alpha\beta}\right)$$

completes the field–theoretic picture: it encodes gravitational energy density, momentum density, and flux; it obeys a local conservation law in conjunction with the matter tensor; and it shows explicitly how the field transmits energy–momentum through space in a manner fully analogous to the electromagnetic Poynting vector.

Thus, Flat Spacetime Gravity achieves what is often considered impossible within a flat geometry: a complete Lagrangian foundation for gravitational interactions, unified in form with electromagnetism, locally conserved, and fully dynamical. This synthesis—linear field equations, Lorentz–type force law, conserved stress–energy, and an effective potential that incorporates the geometry of motion—stands as one of the most striking conceptual features of the theory.

7. Relativistic Hamiltonian Formulation

7.1 Introductory Remark

The transition from a Lagrangian to a Hamiltonian formulation is one of the deep structural tests of any field theory. In electromagnetism, the passage to canonical variables (A_μ, Π^μ) reveals the constraint nature of the scalar potential, the gauge freedom generated by Gauss's law, and the clean separation between dynamical evolution and algebraic restrictions. A flat-spacetime gravitational theory that is genuinely field–theoretic—rather than purely geometric—must be able to sustain the same transition.

The Hamiltonian formulation presented in this section establishes that the flat-spacetime gravitational field admits a complete canonical structure on flat Minkowski spacetime. The basic field variable is the 4–potential Φ_μ, while its conjugate momentum Π^μ is proportional to the sourceful gravitational field. The temporal component Φ_0 naturally emerges as a Lagrange multiplier enforcing the Gauss constraint, and the evolution of the spatial components is governed by a Hamiltonian density quadratic in the invariants $G_{\mu\nu}G^{\mu\nu}$. In close parallel with electromagnetism, the Hamiltonian equations reproduce the full covariant system

$$\partial_\mu G^{\mu\nu} = -\frac{4\pi G}{c^2} J^\nu, \qquad \partial_{[\lambda} G_{\mu\nu]} = 0$$

demonstrating that the flat-spacetime gravitational field is a genuine Minkowskian gauge theory with well–posed canonical dynamics. This closes the conceptual circle begun in the Lagrangian formulation and places the theory on the same mathematical footing as Maxwell electrodynamics.

Beyond establishing the purely classical canonical structure, this section also prepares the transition towards semiclassical and quantum descriptions. The Hamilton–Jacobi formulation is recast in a WKB language, the bound–orbit dynamics of the FSG Kepler problem is encoded in action–angle variables, and the test–particle sector admits a fully canonical quantisation leading to a Klein–Gordon–type wave equation. These results provide the natural bridge between the classical gauge field description and the semiclassical and quantum analyses developed in the following sections.

Thus, the goal of this section is not merely computational. It is to show— rigorously and without any recourse to curved geometry—that the flat-spacetime gravitational field admits a complete Hamiltonian description parallel to Maxwell theory: with canonical momenta, constraints, evolution equations, and a manifestly 4D covariant recovery of the full field equations. In doing so, the theory demonstrates its internal consistency at a foundational level and reveals the same conceptual architecture that makes electromagnetism one of the cleanest and most transparent field theories in physics.

7.2 Canonical momentum and the particle Hamiltonian

7.2.1 Lagrangian

In the Hamiltonian description we start from the coordinate–time Lagrangian of a test particle moving in the effective potential constructed in the previous section:

$$\mathcal{L}(t,\mathbf{x},\mathbf{v}) \;=\; -m_0c^2\sqrt{1-\frac{v^2}{c^2}} \;-\; m_0\,\phi^{(\mathrm{eff})}(t,\mathbf{x}) \;+\; m_0\,\mathbf{A}^{(\mathrm{eff})}(t,\mathbf{x})\cdot\mathbf{v}$$

The term involving $G_{\mu\nu}G^{\mu\nu}$ in the full action does not depend on $\mathbf{v}$ and therefore plays no role in the definition of the particle's canonical momentum.

7.2.2 Canonical momentum

The canonical momentum conjugate to the spatial coordinates is

$$p_i \;\equiv\; \frac{\partial\mathcal{L}}{\partial v_i} \;=\; m_0\gamma\,v_i \;+\; m_0\,A_i^{(\mathrm{eff})}$$

or in vector notation

$$\mathbf{p} \;=\; m_0\gamma\,\mathbf{v} \;+\; m_0\,\mathbf{A}^{(\text{eff})}$$

It is convenient to define the kinetic (or mechanical) momentum

$$\mathbf{\Pi} \;\equiv\; \mathbf{p} - m_0\,\mathbf{A}^{(\text{eff})} \;=\; m_0\gamma\,\mathbf{v} \;=\; m_0\,\mathbf{u}$$

which depends only on the particle's kinematics.

7.2.3 Hamiltonian

The Hamiltonian is obtained in the standard way,

$$H \;= \mathbf{p}\cdot\mathbf{v} \;-\; \mathcal{L}$$

Using $\mathbf{p} = m_0\gamma\mathbf{v} + m_0\mathbf{A}^{(\text{eff})}$ and the explicit form of $\mathcal{L}$, a short computation yields

$$H \;=\; \left(m_0\gamma\mathbf{v} +\; m_0\,\mathbf{A}^{(\text{eff})}(t,\mathbf{x})\right)\cdot\mathbf{v} +$$

$$m_0c^2\sqrt{1-\frac{v^2}{c^2}} \;+\; m_0\,\phi^{(\text{eff})}(t,\mathbf{x}) \;-\; m_0\,\mathbf{A}^{(\text{eff})}(t,\mathbf{x})\cdot\mathbf{v}$$

so

$$H \;=\; m_0\gamma c^2 + m_0\,\phi^{(\text{eff})}$$

Thus, the Hamiltonian separates cleanly into kinetic and potential pieces:

$$H \;\;=\; m_0\gamma c^2 + m_0\,\phi^{(\text{eff})}$$

completely analogous to the electromagnetic expression $H_{\text{EM}} = \gamma mc^2 + q\phi$.

7.2.4 Constant of motion and the energy parameter *E*

When the effective potential is time–independent, $\partial_t \phi^{(\text{eff})} = 0$, $\partial_t \mathbf{A}^{(\text{eff})} = 0$, the Hamiltonian is conserved along the dynamical trajectory:

$$H = \text{const}$$

It is natural to parameterize this constant in the form

$$H = m_0 c^2 + m_0 E$$

where E carries the interpretation of the conserved energy per unit rest mass (including all effective flat-spacetime gravitational contributions). Substituting this into the previous expression gives

$$m_0 \gamma c^2 + m_0 \phi^{(\text{eff})} = m_0 c^2 + m_0 E$$

and hence

$$\gamma = 1 - \frac{\phi^{(\text{eff})}}{c^2} + \frac{E}{c^2}$$

The above equations complete the Hamiltonian characterization of a test particle moving in the effective flat-spacetime gravitational potential.

7.3 Hamilton's equations and recovery of the effective force law

7.3.1 Hamilton's equations

Having established the canonical momentum and Hamiltonian for a test particle, we now derive the full set of Hamilton's equations and show explicitly that they reproduce the dynamical law

$$a^\mu = G^{(\text{eff})\mu}_{\ \ \nu}\, U^\nu$$

obtained from the Euler–Lagrange formulation in the previous section. This provides a second, independent verification of the consistency of the theory and completes the parallelism with electromagnetism.

The canonical variables are $(\mathbf{x}, \mathbf{p})$ with Hamiltonian

$$H(\mathbf{x}, \mathbf{p}) \;\; = \; m_0 \gamma(\mathbf{\Pi})\, c^2 + m_0\, \phi^{(\text{eff})}(\mathbf{x})$$

where the kinetic (mechanical) momentum is

$$\mathbf{\Pi} \; = \; \mathbf{p} - m_0\, \mathbf{A}^{(\text{eff})}(\mathbf{x}), \qquad\qquad \gamma \; = \; \sqrt{1 + \frac{\Pi^2}{m_0^2 c^2}}$$

Hamilton's equations read

$$\dot{x}^i \; = \; \frac{\partial H}{\partial p_i}, \qquad\qquad \dot{p}_i \;\; = \; -\frac{\partial H}{\partial x^i}$$

where the overdot denotes differentiation with respect to coordinate time t.

7.3.2 First Hamilton equation

Using

$$\frac{\partial \gamma}{\partial \Pi_i} \; = \; \frac{\Pi_i}{m_0^2 c^2 \gamma} \; = \; \frac{v_i}{c^2}$$

and $\partial \Pi_j / \partial p_i \; = \; \delta_{ij}$, we obtain

$$\dot{x}^i = c^2 m_0 \frac{\partial \gamma}{\partial p_i} = c^2 m_0 \frac{\partial \gamma}{\partial \Pi_i} = \frac{\Pi_i}{\gamma m_0} = v_i$$

Thus, the first Hamilton equation trivially reproduces the kinematic identity

$$\dot{\mathbf{x}} \; = \; \mathbf{v}$$

7.3.3 Second Hamilton equation

The position dependence of H enters through $\phi^{(\text{eff})}(\mathbf{x})$ and $\mathbf{A}^{(\text{eff})}(\mathbf{x})$. We compute

$$\dot{p}_i \;=\; -\frac{\partial H}{\partial x^i} \;=\; -m_0\,\partial_i\phi^{(\text{eff})} \;-\; c^2 m_0 \frac{\partial\gamma}{\partial\Pi_k}\frac{\partial\Pi_k}{\partial x^i}$$

Since

$$\frac{\partial\gamma}{\partial\Pi_k} = \frac{v_k}{c^2}, \qquad\qquad \partial_i\Pi_k \;=\; -m_0\,\partial_i A_k^{(\text{eff})}$$

we obtain

$$\dot{p}_i \;=\; -m_0\,\partial_i\phi^{(\text{eff})} + m_0\,v_k\,\partial_i A_k^{(\text{eff})}$$

But

$$\dot{\Pi}_i \;=\; \dot{p}_i - m_0\frac{d}{dt}A_i^{(\text{eff})} \;=\; \dot{p}_i - m_0\left(\partial_t A_i^{(\text{eff})} + v_j\,\partial_j A_i^{(\text{eff})}\right)$$

so, substituting the expression for $\dot{p}_i$ yields

$$\dot{\Pi}_i \;=\; -m_0\,\partial_i\phi^{(\text{eff})} + m_0\,v_k\,\partial_i A_k^{(\text{eff})} - m_0\,\partial_t A_i^{(\text{eff})} - m_0\,v_j\,\partial_j A_i^{(\text{eff})}$$

Grouping terms and using

$$(\mathbf{v}\times\mathbf{b}^{(\text{eff})})_i \;=\; v_j\left(\partial_i A_j^{(\text{eff})} - \partial_j A_i^{(\text{eff})}\right)$$
$$\mathbf{g}^{(\text{eff})} \;=\; -\nabla\phi^{(\text{eff})} - \partial_t\mathbf{A}^{(\text{eff})}$$

we obtain

$$m_0\,\dot{\mathbf{u}} \;=\; m_0\,\mathbf{g}^{(\text{eff})} + m_0\,\mathbf{v}\times\mathbf{b}^{(\text{eff})}$$

or equivalently:

$$\dot{\mathbf{u}} \;=\; \mathbf{g}^{(\text{eff})} + \mathbf{v}\times\mathbf{b}^{(\text{eff})}$$

7.3.4 Covariant form

Writing $U^\mu = (c\gamma, \gamma v^i)$ and using the definitions

$$G_{0i}^{(\text{eff})} = -\frac{1}{c}\, g_i^{(\text{eff})}$$
$$G_{ij}^{(\text{eff})} = \epsilon_{ijk}\, b_k^{(\text{eff})}$$

The previous equation is exactly the spatial part of

$$a^\mu = \mathrm{G}_\nu^{(\text{eff})\mu}\, U^\nu$$

the dynamical law derived previously from the Euler–Lagrange formulation.

7.3.5 Conclusion

Hamilton's equations reproduce the full effective flat-spacetime gravitational force law with no additional assumptions. The equivalence between the Lagrangian and Hamiltonian formulations is therefore complete, and the analogy with electromagnetism extends not only to the field equations, but also to the particle dynamics and the canonical phase–space structure.

7.4 Gauge structure and canonical equivalence

7.4.1 Gauge transformation of the canonical momentum

A central feature of the Flat Spacetime Gravity is that the fundamental tensor

$$G_{\mu\nu}^{(\text{eff})} = \partial_\mu \Phi_\nu^{(\text{eff})} - \partial_\nu \Phi_\mu^{(\text{eff})}$$

is invariant under the gauge transformation

$$\Phi_\mu^{(\text{eff})} \rightarrow \Phi_\mu^{(\text{eff})} + \partial_\mu \Lambda, \qquad \Lambda = \Lambda(x^\alpha)$$

for any smooth scalar function Λ. This mirrors the structure of electromagnetism, but is here realised in purely flat spacetime without curvature.

Under a gauge transformation the coordinate–time Lagrangian changes by a total derivative:

$$\mathcal{L} \;\to\; \mathcal{L}' \;=\; \mathcal{L} + m_0 \frac{d\Lambda}{dt}$$

Since total time derivatives do not affect the equations of motion, the dynamics remains unchanged. The canonical momentum transforms as

$$p_i \;=\; \frac{\partial \mathcal{L}}{\partial v_i} \qquad \to \qquad p_{i'} \;=\; p_i + m_0\, \partial_i \Lambda$$

while the mechanical momentum

$$\mathbf{\Pi} \;=\; \mathbf{p} - m_0\, \mathbf{A}^{(\mathrm{eff})}$$

is invariant:

$$\mathbf{\Pi}' \;=\; \mathbf{p}' - m_0\big(\mathbf{A}^{(\mathrm{eff})} + \nabla\Lambda\big) \;=\; \mathbf{p} - m_0\, \mathbf{A}^{(\mathrm{eff})} \;=\; \mathbf{\Pi}$$

Therefore, all kinetic quantities, including γ and U^{μ}, are gauge–independent, exactly as in electromagnetism.

7.4.2 Gauge transformation of the Hamiltonian

Because the Hamiltonian depends on **x** and **p** through

$$H(\mathbf{x},\mathbf{p}) \;=\; m_0\gamma(\mathbf{\Pi})\, c^2 \;+\; m_0\, \phi^{(\mathrm{eff})}$$

the gauge shift modifies H exactly as

$$H' \;=\; H + m_0\, \partial_t \Lambda$$

But the canonical transformation

$$\begin{aligned} \mathbf{p} \;\to\; \mathbf{p}' \;&=\; \mathbf{p} + m_0 \nabla\Lambda \\ H \;\to\; H' \;&=\; H + m_0\, \partial_t \Lambda \end{aligned}$$

preserves Hamilton's equations. Thus, the gauge transformation is a legitimate canonical transformation in phase space.

7.4.3 Invariance of the equations of motion

The physically observable fields

$$\begin{aligned} \mathbf{g}^{(\text{eff})} &= -\nabla\phi^{(\text{eff})} - \partial_t\mathbf{A}^{(\text{eff})} \\ \mathbf{b}^{(\text{eff})} &= \nabla\times\mathbf{A}^{(\text{eff})} \end{aligned}$$

remain unchanged under

$$\begin{aligned} \phi^{(\text{eff})} &\to \phi^{(\text{eff})} + \partial_t\Lambda \\ \mathbf{A}^{(\text{eff})} &\to \mathbf{A}^{(\text{eff})} + \nabla\Lambda \end{aligned}$$

Consequently, the dynamical law

$$\dot{\mathbf{u}} = \mathbf{g}^{(\text{eff})} + \mathbf{v}\times\mathbf{b}^{(\text{eff})}$$

and the covariant expression

$$a^{\mu} = \mathrm{G}_{\nu}^{(\text{eff})\mu} U^{\nu}$$

are strictly gauge–invariant.

7.4.4 Interpretation

Thus, the gauge freedom of $\Phi_{\mu}^{(\text{eff})}$ acts exactly as the electromagnetic gauge freedom but with an important conceptual distinction: the flat-spacetime gravitational theory is entirely flat, linear, and Minkowskian, yet possesses a nontrivial gauge symmetry whose invariants encode all flat-spacetime gravitational effects. This completes the parallel with electromagnetism at the Hamiltonian level and reinforces the internal consistency of the effective potential formalism introduced in previous sections.

7.5 Hamilton–Jacobi formulation

7.5.1 Hamilton–Jacobi equation

The Hamilton–Jacobi formalism provides an alternative, fully covariant description of particle dynamics in the effective flat-spacetime gravitational potential. Its importance is twofold: (i) it yields a scalar action function whose level surfaces encode the entire trajectory, and (ii) it prepares the ground for semiclassical, wave–mechanical or eikonal interpretations, exactly as in electromagnetism and standard relativistic mechanics.

The Hamiltonian obtained previously,

$$H(\mathbf{x}, \mathbf{p}) \;=\; m_0\gamma(\mathbf{\Pi})\, c^2 \;+\; m_0\, \phi^{(\mathrm{eff})}(\mathbf{x})$$

with

$$\mathbf{\Pi} \;= \mathbf{p} - m_0\, \mathbf{A}^{(\mathrm{eff})}(\mathbf{x}), \qquad \gamma = \sqrt{1 + \frac{\Pi^2}{m_0^2 c^2}}$$

leads directly to the Hamilton–Jacobi equation by the standard substitution

$$\mathbf{p} \;\rightarrow\; \nabla S, \qquad H(\mathbf{x}, \nabla S) + \frac{\partial S}{\partial t} = 0$$

where $S(\mathbf{x}, t)$ is Hamilton's principal function.

Thus, the Hamilton–Jacobi equation for a massive particle in the effective potential is

$$\frac{\partial S}{\partial t} + m_0 c^2 \sqrt{1 + \frac{(\nabla S - m_0\, \mathbf{A}^{(\mathrm{eff})})^2}{m_0^2 c^2}} + m_0\, \phi^{(\mathrm{eff})} \;=\; 0$$

7.5.2 Gauge transformation of the Hamilton–Jacobi function

A gauge transformation

$$\Phi_{\mu}^{(\text{eff})} \rightarrow \Phi_{\mu}^{(\text{eff})} + \partial_{\mu}\Lambda$$

implies

$$\phi^{(\text{eff})} \rightarrow \phi^{(\text{eff})} + \partial_{t}\Lambda, \qquad \mathbf{A}^{(\text{eff})} \rightarrow \mathbf{A}^{(\text{eff})} + \nabla\Lambda$$

To preserve the Hamilton–Jacobi equation, the principal function must transform as

$$S \rightarrow S' = S + m_0\Lambda$$

for which

$$\nabla S' - m_0\left(\mathbf{A}^{(\text{eff})} + \nabla\Lambda\right) = \nabla S - m_0\,\mathbf{A}^{(\text{eff})}$$

and likewise

$$\frac{\partial S'}{\partial t} + m_0\,\partial_t\Lambda = \frac{\partial S}{\partial t}$$

Thus, the Hamilton–Jacobi equation is strictly gauge–invariant, reflecting the gauge freedom intrinsic to $\Phi_{\mu}^{(\text{eff})}$.

7.5.3 Recovery of the dynamical trajectory

Differentiating the principal function yields the canonical momentum,

$$p_i = \frac{\partial S}{\partial x^i}$$

and hence the mechanical momentum,

$$\Pi_i = p_i - m_0 A_i^{(\text{eff})} = m_0\gamma v_i$$

Therefore the 4–velocity is reconstructed from S via

$$U^i = \gamma v^i = \frac{1}{m_0}\left(\frac{\partial S}{\partial x^i} - m_0 A_i^{(\mathrm{eff})}\right), \qquad U^0 = c\gamma$$

The Hamilton–Jacobi function S thus encodes the full worldline of the particle; its level surfaces determine the flow of U^μ, and the corresponding trajectory automatically satisfies the effective-force law

$$a^\mu = \mathrm{G}_\nu^{(\mathrm{eff})\mu} U^\nu$$

derived previously from both the Euler–Lagrange and Hamiltonian formulations.

7.5.4 Interpretation

The Hamilton–Jacobi formulation provides a scalar, gauge–covariant, and geometrically transparent characterization of particle motion in the flat-spacetime gravitational field. Its structural identity with the electromagnetic Hamilton–Jacobi equation further highlights the deep analogy between the two theories, even though the present framework remains entirely Minkowskian and flat. The function S may also serve as the starting point for semiclassical or eikonal approximations, where wavefronts propagate according to the effective potential $\Phi_\mu^{(\mathrm{eff})}$, completing yet another parallel with electrodynamics.

7.6 Reparameterization constraint

7.6.1 Reparameterization invariance of the particle action

A distinctive feature of relativistic particle dynamics is the invariance of the covariant action under arbitrary reparameterizations of the worldline. This symmetry implies the presence of a primary constraint among the canonical phase–space variables (x^μ, p_μ). In the flat-spacetime gravitational theory the structure is fully analogous to electromagnetism, but realised on a strictly flat Minkowski background.

The covariant particle action is

$$S_{\text{part}} = \int \left(-m_0 c^2 + m_0\, \Phi_\mu^{(\text{eff})}\, U^\mu\right) d\tau$$

where $U^\mu = dx^\mu/d\tau$. Under any monotonic change of parameter,

$$\tau \;\rightarrow\; \tau' = f(\tau), \qquad \frac{df}{d\tau} > 0$$

the action remains invariant. This symmetry implies, by Dirac's theory of constrained Hamiltonian systems, a primary constraint relating the canonical variables.

7.6.2 Canonical momentum

From the covariant Lagrangian

$$L = -m_0 c^2 + m_0\, \Phi_\mu^{(\text{eff})}\, U^\mu$$

the canonical momentum is

$$p_\mu \equiv \frac{\partial L}{\partial U^\mu} = m_0\, U_\mu + m_0\, \Phi_\mu^{(\text{eff})}$$

Solving for the 4–velocity,

$$U_\mu = \frac{1}{m_0}\left(p_\mu - m_0\, \Phi_\mu^{(\text{eff})}\right)$$

7.6.3 Mass–shell constraint

The normalization of the 4–velocity,

$$U_\mu U^\mu = -c^2$$

now implies

$$\frac{1}{m_0^2}\left(p_\mu - m_0 \Phi_\mu^{(\text{eff})}\right)\left(p^\mu - m_0 \Phi^{(\text{eff})\mu}\right) = -\,c^2$$

Multiplying through by m_0^2 yields the primary constraint

$$\left(p_\mu - m_0\Phi_\mu^{(\text{eff})}\right)\left(p^\mu - m_0\Phi^{(\text{eff})\mu}\right) + m_0^2c^2 = 0$$

This is the flat-spacetime gravitational analogue of the electromagnetic mass–shell condition

$$(p_\mu - qA_\mu)(p^\mu - qA^\mu) + m_0^2c^2 = 0$$

with the effective potential playing the role of the gauge connection

7.6.4 Relation to the coordinate–time Hamiltonian

In previous subsections we fixed the worldline parameter by choosing $t = x^0/c$, breaking explicitly the reparametrisation symmetry and producing a true Hamiltonian

$$H = m_0\gamma c^2 + m_0\,\phi^{(\text{eff})} = m_0c^2 + m_0E$$

To connect this with the covariant constraint given before, note that

$$p_0 = \frac{\partial L}{\partial U^0} = m_0U_0 + m_0\,\Phi_0^{(\text{eff})}$$

and solving the constraint for p_0 gives

$$p_0 = -H + m_0\,\phi^{(\text{eff})}$$

which is precisely the Legendre transform relation

$$p_0 = -\frac{\partial S}{\partial t}$$

Thus, the coordinate–time Hamiltonian arises as the solution of the primary constraint with respect to p_0, confirming the mathematical equivalence of the reduced and fully covariant formalisms.

7.6.5 Interpretation

The constraint introduced here expresses the fact that the canonical variables live not in an 8–dimensional phase space but on a 7–

dimensional constraint surface, as required by reparameterization invariance. The structure is identical to that of electromagnetism but here realized in a strictly flat Minkowski spacetime, fully consistent with the general philosophy of the Flat Spacetime Gravity.

Most importantly, the constraint guarantees that the phase–space variables always reconstruct a valid 4–velocity and that the Hamiltonian equations of motion are dynamically equivalent to the Euler–Lagrange equations derived in the Lagrangian section.

7.7 Canonical field variables and the Legendre transformation

7.7.1 Canonical momenta

Before deriving the Hamiltonian field equations and the Maxwell–type structure of the flat-spacetime gravitational theory, we must construct the canonical variables associated with the flat-spacetime gravitational potential Φ_μ. The construction follows the same steps as in electromagnetism but with the specific tensor structure of

$$G_{\mu\nu} = \partial_\mu \Phi_\nu - \partial_\nu \Phi_\mu$$

and with the flat–space Lagrangian density

$$\mathcal{L}_{\text{field}} = -\frac{c^2}{16\pi G} G_{\mu\nu} G^{\mu\nu}$$

Only the “bare” potential Φ_μ enters here: the orbital part $\Phi_\mu^{(\text{orb})}$ is *not* varied at the field level, in full agreement with the structure established in Lagrangian section.

Since the field variables are the four components Φ_μ, the canonical momenta conjugate to Φ_μ are defined by the functional derivative

$$\Pi^\mu \equiv \frac{\partial \mathcal{L}_{\text{field}}}{\partial(\partial_t \Phi_\mu)}$$

Using

$$G_{0\mu} = \partial_0 \Phi_\mu - \partial_\mu \Phi_0, \qquad \partial_0 \equiv \frac{\partial}{\partial t}$$

and noting that only $G_{0\mu}$ depends on the time derivative of Φ_μ, we obtain

$$\Pi^\mu = -\frac{c^2}{8\pi G} G^{0\mu}$$

or explicitly

$$\Pi^0 = 0, \qquad \mathbf{\Pi} = -\frac{c^2}{8\pi G} \mathbf{g}$$

where $\mathbf{g} = (g_1, g_2, g_3)$ are the components of the sourceful gravitational field.

Equation above shows that Φ_0 has no time derivatives in the Lagrangian; consequently,

$$\Pi^0 = 0$$

is a *primary constraint* of the field theory, completely analogous to the vanishing canonical momentum conjugate to the electromagnetic scalar potential A_0.

7.7.2 Velocity–momentum relation

The time derivative of the spatial potential is encoded in G_{0i}:

$$G_{0i} = \partial_t \Phi_i - \partial_i \Phi_0 = -\frac{8\pi G}{c^2} \Pi^i$$

Solving for $\partial_t \Phi_i$ gives

$$\partial_t \Phi_i = -\frac{8\pi G}{c^2} \Pi^i + \partial_i \Phi_0$$

7.7.3 Legendre transform and field Hamiltonian density

The field Hamiltonian density is defined as

$$\mathcal{H}_{\text{field}} \equiv \Pi^{\mu}\, \partial_t \Phi_{\mu} - \mathcal{L}_{\text{field}}$$

Using $\Pi^0 = 0$ and the velocity–momentum relation,

$$\Pi^i\, \partial_t \Phi_i = \Pi^i \left(-\frac{8\pi G}{c^2} \Pi^i + \partial_i \Phi_0 \right)$$

and the identity

$$G_{\mu\nu} G^{\mu\nu} = 2\left(\mathbf{b}^2 - \frac{\mathbf{g}^2}{c^2} \right)$$

we obtain after straightforward algebra

$$\mathcal{H}_{\text{field}} = \frac{c^2}{16\pi G} (\mathbf{g}^2 + \mathbf{b}^2) + \Phi_0\, (\nabla \cdot \mathbf{g})$$

Here the last term is proportional to the Gauss constraint; in the full theory it becomes $\Phi_0\,(-4\pi G \rho_0 \gamma)$ after coupling to matter. Thus Φ_0 acts as a Lagrange multiplier enforcing the field constraint, just as the scalar potential A_0 enforces Gauss's law in electromagnetism.

7.7.4 Interpretation

Two essential structural features emerge:

- $\Pi^0 = 0$ is a primary constraint expressing the fact that the scalar potential Φ_0 is not a dynamical variable. It plays the same role as the temporal component of the gauge field in electromagnetism.

- The canonical spatial momentum is proportional to the sourceful gravitational field $\mathbf{g}$:

$$\mathbf{\Pi} = -\frac{c^2}{8\pi G} \mathbf{g}$$

  Thus, the sourceful gravitational field $\mathbf{g}$ is the canonical momentum density, while the rotational gravitational field $\mathbf{b}$

appears algebraically in the Hamiltonian, exactly mirroring electrodynamics.

These results set the stage for the derivation of the Hamiltonian field equations in the next subsections.

7.8 Hamiltonian field equations

7.8.1 Canonical momenta and Hamiltonian density

In previous section we obtained the Maxwell–type field equations for the bare flat-spacetime gravitational field by varying the action with respect to the potential Φ_μ. The result can be written in covariant form as

$$\partial_\mu G^{\mu\nu} = -\frac{4\pi G}{c^2} J^\nu$$

together with the Bianchi identity

$$\partial_\lambda G_{\mu\nu} + \partial_\mu G_{\nu\lambda} + \partial_\nu G_{\lambda\mu} = 0$$

In this subsection we show that the same equations follow from the Hamiltonian field formulation built on the canonical variables (Φ_μ, Π^μ).

The field Lagrangian density is

$$\mathcal{L}_{\text{field}} = -\frac{c^2}{16\pi G} G_{\mu\nu} G^{\mu\nu}, \qquad G_{\mu\nu} = \partial_\mu \Phi_\nu - \partial_\nu \Phi_\mu$$

The canonical momenta are

$$\Pi^\mu \equiv \frac{\partial \mathcal{L}_{\text{field}}}{\partial(\partial_t \Phi_\mu)} = -\frac{c^2}{8\pi G} G^{0\mu}$$

so that

$$\Pi^0 = 0, \qquad \mathbf{\Pi} \propto \mathbf{g}$$

with $\mathbf{g}$ the sourceful gravitational field. The scalar potential Φ_0 is therefore non–dynamical and will act as a Lagrange multiplier enforcing the Gauss constraint.

Including the matter coupling, the total field Hamiltonian density can be written in the schematic form

$$\mathcal{H} = \mathcal{H}_{\text{kin}}(\mathbf{\Pi}, \mathbf{b}) + \Phi_0 \mathcal{G} - \frac{1}{c^2} \Phi_\nu J^\nu$$

where $\mathcal{H}_{\text{kin}}$ is quadratic in $\mathbf{g}$ and $\mathbf{b}$, and $\mathcal{G}$ is the Gauss constraint (a spatial divergence of $\mathbf{g}$). The explicit coefficients are chosen so that this equation follows with exactly the same normalization as in Lagrangian section.

7.8.2 Hamilton's equations for Φ_μ

Hamilton's equations for fields read

$$\partial_t \Phi_\mu = \frac{\delta \mathcal{H}}{\delta \Pi^\mu}, \qquad \partial_t \Pi^\mu = -\frac{\delta \mathcal{H}}{\delta \Phi_\mu}$$

For the temporal component we have

$$\partial_t \Phi_0 = \frac{\delta \mathcal{H}}{\delta \Pi^0} = 0$$

which confirms that Φ_0 is a non–dynamical multiplier.

For the spatial components one finds, using the definition of Π^i in terms of G^{0i} and hence $\mathbf{g}$,

$$\partial_t \Phi_i = \frac{\delta \mathcal{H}}{\delta \Pi^i} = (\text{expression proportional to } G_{0i}) = \partial_t \Phi_i - \partial_i \Phi_0$$

which is nothing but the definition of G_{0i} written in Hamiltonian form. Thus, the first set of Hamiltonian equations reproduces the relation

$$G_{0i} = \partial_t \Phi_i - \partial_i \Phi_0$$

7.8.3 Hamilton's equations for Π^μ

The second set of Hamilton equations gives the evolution and constraints.

Temporal component

Varying $\mathcal{H}$ with respect to Φ_0 yields

$$\partial_t \Pi^0 = -\frac{\delta \mathcal{H}}{\delta \Phi_0} = -\mathcal{G} + \frac{1}{c^2} J^0$$

Since $\Pi^0 = 0$ identically, we obtain the constraint

$$\mathcal{G} = \frac{1}{c^2} J^0$$

With the explicit identification of $\mathcal{G}$ as the spatial divergence of the sourceful gravitational field $\mathbf{g}$, this is exactly the $\nu = 0$ component of the previous equation:

$$\partial_\mu G^{\mu 0} = -\frac{4\pi G}{c^2} J^0$$

Spatial components

Variation with respect to Φ_i gives

$$\partial_t \Pi^i = -\frac{\delta \mathcal{H}}{\delta \Phi_i} = (\nabla \times \mathbf{b})^i + \frac{4\pi G}{c^2} J^i$$

thus

$$(\nabla \times \mathbf{b})_i = -\frac{4\pi G}{c^2} J_i + \frac{1}{c^2} \frac{\partial g_i}{\partial t}$$

Using again the relation between $\mathbf{\Pi}$ and G^{0i}, this is equivalent to the spatial components of the previous equation:

$$\partial_\mu G^{\mu i} = -\frac{4\pi G}{c^2} J^i$$

Covariant reconstruction

Collecting the temporal and spatial components, we see that the Hamiltonian field equations reproduce the full inhomogeneous Maxwell–type system

$$\partial_\mu G^{\mu\nu} = -\frac{4\pi G}{c^2} J^\nu$$

with the Bianchi identity still following directly from the definition of $G_{\mu\nu}$ in terms of Φ_μ.

7.8.4 Conclusion

The Hamiltonian formulation is therefore fully equivalent to the Lagrangian formulation of previous section. The canonical structure based on (Φ_μ, Π^μ) leads to exactly the same Maxwell–type equations with the correct normalisation $-4\pi G/c^2$, confirming that the flat-spacetime gravitational field theory admits a consistent Hamiltonian description in which constraints, gauge freedom, and dynamical evolution are all handled in complete analogy with electromagnetism.

7.9 WKB Approximation and the Classical Limit

7.9.1 WKB ansatz

The Hamilton–Jacobi equation obtained in the previous subsection admits a natural semiclassical interpretation through the WKB method. As in electromagnetism, the central idea is that a wavefunction whose phase varies rapidly compared to its amplitude behaves, in the limit $\hbar \to 0$, as a classical particle following the characteristics of the Hamilton–Jacobi equation. This subsection develops that correspondence for the FSG dynamics.

Consider a scalar wavefunction of the form

$$\psi(x) = A(x)\exp\left(\frac{i}{\hbar}S(x)\right)$$

where S is the rapidly varying phase and A is a slowly varying amplitude. Substituting this equation into the minimally coupled wave equation

$$\left[(-i\hbar\,\partial_\mu - m_0\Phi_\mu^{(\mathrm{eff})})(-i\hbar\,\partial^\mu - m_0\Phi^{(\mathrm{eff})\mu}) + m_0^2c^2\right]\psi = 0$$

and collecting powers of $\hbar$, we obtain at leading order $\hbar^0$:

$$\left(\partial_\mu S - m_0\Phi_\mu^{(\mathrm{eff})}\right)\left(\partial^\mu S - m_0\Phi^{(\mathrm{eff})\mu}\right) = -m_0^2c^2$$

This is precisely the Hamilton–Jacobi equation we have already seen. Thus, the phase S of a WKB wavefunction must satisfy the same gauge–covariant constraint as the classical action of a test particle.

7.9.2 Emergence of the classical 4–velocity

At leading WKB order the wave has a locally defined momentum

$$p_\mu = \partial_\mu S, \qquad \Pi_\mu = p_\mu - m_0\Phi_\mu^{(\mathrm{eff})}$$

and the 4–velocity emerges automatically as

$$U_\mu = \frac{\Pi_\mu}{m_0}$$

Hence the rays of the WKB approximation coincide with the particle worldlines generated by the FSG dynamical law. This reproduces the classical relation

$$a^\mu = \mathrm{G}_\nu^{(\mathrm{eff})\mu}U^\nu$$

showing that the classical motion is the geometric trace of the underlying WKB phase evolution.

7.9.3 Transport of the amplitude

The next order in $\hbar$ produces a transport equation for the amplitude A:

$$\partial_\mu(A^2\Pi^\mu) \;=\; 0$$

expressing probability (or flux) conservation along the characteristics. The structure is again precisely analogous to electromagnetism: the effective momentum Π^μ controls both the phase and the transport of amplitude.

7.9.4 Interpretation and significance

The WKB limit provides a direct bridge between the quantum and classical descriptions. A wave packet in the Flat Spacetime Gravity field propagates along trajectories that satisfy the classical dynamical equation derived in previous sections. No curved metric or geodesics are involved: the entire mechanism operates within flat Minkowski space through the gauge–covariant combination $\partial_\mu S - m_0\Phi_\mu^{(\mathrm{eff})}$.

Thus, the WKB approximation confirms the internal consistency of the theory: the classical FSG dynamics emerges from a quantum wave equation by the same mechanism through which the Lorentz force emerges from the WKB limit of electromagnetism. This prepares the ground for the canonical quantisation of the particle and the semiclassical action–angle analysis developed in the next subsections.

7.10 Action–Angle Variables for the FSG Kepler Problem

7.10.1 Separation of variables

The Hamilton–Jacobi formulation developed in the previous subsection enables a canonical action–angle description of bounded orbits in the Flat Spacetime Gravity field. For central configurations the effective potential depends only on r, and the motion separates into radial and angular components in the standard way. The purpose of this subsection is to show that the complete FSG dynamics of a bound orbit can be encoded in two invariant actions, (J_r, J_ϕ), whose explicit forms reproduce the classical results of the previous sections, including the perihelion advance.

For planar motion in a central potential the Hamilton–Jacobi function admits a separable ansatz

$$S(t,r,\phi) \;=\; -Ht + S_r(r) + \ell\,\phi$$

where $H = m_0c^2 + m_0E$ is the conserved energy and ℓ is the conserved angular momentum. The effective FSG potential consists of two parts,

$$\phi^{(\mathrm{eff})}(r) \;=\; \phi(r) + \phi^{(\mathrm{orb})}(r) \;=\; -\frac{GM}{r} - \frac{GMh^2}{c^2r^3}$$

with $h = |\mathbf{r} \times \mathbf{u}|$ the specific angular momentum. Inserting this structure into the Hamilton–Jacobi equation yields

$$\left(\frac{dS_r}{dr}\right)^2 \;=\; 2m_0^2\left[E - \phi^{(\mathrm{eff})}(r)\right] - \frac{\ell^2}{r^2}$$

7.10.2 Definition of the action variables

The canonical actions are defined as

$$J_\phi \;=\; \frac{1}{2\pi}\oint p_\phi\, d\phi \;=\; \ell$$

$$\begin{aligned} J_r \;&=\; \frac{1}{2\pi}\oint p_r\, dr \\ &=\; \frac{1}{\pi}\int_{r_{\min}}^{r_{\max}} \sqrt{2m_0^2[E - \phi^{(\mathrm{eff})}(r)] - \frac{\ell^2}{r^2}}\; dr \end{aligned}$$

These integrals contain the entire dynamical information of the bound orbit. The turning points $r_{\min}$ and $r_{\max}$ are the roots of the expression under the square root in the previous equation.

7.10.3 Evaluation of J_r and perihelion advance

To first order in GM/c^2, the radial action can be computed explicitly:

$$J_r = J_r^{(0)} - \frac{3GMm_0^2 h}{c^2}$$

where $J_r^{(0)}$ is the Newtonian action. The angular frequency associated with ϕ is

$$\omega_\phi = \frac{\partial H}{\partial J_\phi} = \omega_0 + \frac{3GM}{c^2 a(1-e^2)}$$

and the radial frequency is

$$\omega_r = \frac{\partial H}{\partial J_r} = \omega_0$$

The difference of frequencies produces the perihelion advance per orbit:

$$\Delta\phi = 2\pi\left(\frac{\omega_\phi}{\omega_r} - 1\right) = \frac{6\pi GM}{c^2 a(1-e^2)}$$

in exact agreement with the dynamical derivation of the perihelion advance of the previous sections.

7.10.4 Interpretation

The action–angle description provides an alternative, canonical route to the same physical predictions obtained earlier. The entire FSG correction to Newtonian gravity is encoded in the shifted radial action J_r, which captures the influence of the orbital potential $\phi^{(\mathrm{orb})}$. No curved metric or geodesic deviation is required: the perihelion advance emerges as a pure action shift in a flat–space Hamiltonian system with a Maxwell–type effective potential.

7.10.5 Significance

The FSG Kepler problem therefore admits a fully integrable action–angle structure, consistent with its Hamilton–Jacobi formulation. This duality between the dynamical law, the HJ method, and the action–angle picture demonstrates that the Flat Spacetime Gravity field possesses the same canonical richness as the electromagnetic interaction. The stage is now set for the semiclassical and quantum analyses developed in the subsequent subsections.

7.11 Canonical Quantisation of the Test Particle

7.11.1 Motivation and strategy

The Hamilton–Jacobi and WKB analyses of the previous subsections show that a test particle in the Flat Spacetime Gravity field behaves exactly like a particle minimally coupled to a Maxwell–type 4–potential. This structural analogy allows a fully canonical quantisation procedure, parallel to the standard quantisation of a charged particle in electromagnetism. The key step is to promote the canonical momenta to operators and impose the corresponding wave equation consistent with the mass–shell constraint.

7.11.2 Canonical momentum and operator substitution

From the Hamiltonian section the canonical momentum is

$$p_\mu \;=\; m_0 U_\mu + m_0\, \Phi_\mu^{(\mathrm{eff})}$$

Canonical quantisation prescribes the replacement

$$p_\mu \;\to\; \hat{p}_\mu \;=\; i\hbar\, \partial_\mu$$

so that the gauge–covariant (kinematical) momentum becomes the operator

$$\hat{\Pi}_\mu = \hat{p}_\mu - m_0\Phi_\mu^{(\text{eff})} = -i\hbar\,\partial_\mu - m_0\Phi_\mu^{(\text{eff})}$$

7.11.3 Quantum wave equation

The classical mass–shell relation

$$\Pi_\mu \Pi^\mu = -m_0^2 c^2$$

is promoted to the operator equation

$$\left[\left(-i\hbar\,\partial_\mu - m_0\Phi_\mu^{(\text{eff})}\right)\left(-i\hbar\,\partial^\mu - m_0\Phi^{(\text{eff})\mu}\right) + m_0^2c^2\right]\psi = 0$$

which is the natural analogue of the Klein–Gordon equation with minimal coupling. This equation is gauge covariant under

$$\Phi_\mu^{(\text{eff})} \mapsto \Phi_\mu^{(\text{eff})} + \partial_\mu\Lambda$$
$$\psi \mapsto \exp\left(\frac{im_0}{\hbar}\Lambda\right)\psi$$

confirming that the FSG interaction is introduced through a consistent gauge mechanism entirely parallel to electromagnetism.

7.11.4 Time–dependent form and nonrelativistic limit

Writing

$$\psi = e^{-\frac{iHt}{\hbar}}\chi(\mathbf{x})$$

and expanding this equation for small velocities, one obtains the nonrelativistic wave equation

$$i\hbar\frac{\partial\chi}{\partial t} = \left[m_0\phi^{(\text{eff})} + \frac{1}{2m_0}(-i\hbar\nabla - m_0\mathbf{A}^{(\text{eff})})^2\right]\chi$$

which is the Schrödinger–type equation associated with the effective FSG potential. The Newtonian potential and the orbital contribution enter symmetrically through $\phi^{(\text{eff})}$, while the gravito-rotational interaction enters through $\mathbf{A}^{(\mathrm{eff})}$ in strict analogy with the minimal coupling of electromagnetism.

7.11.5 Interpretation

The canonical quantisation of the FSG particle is therefore completely well–defined within flat Minkowski spacetime. The structure mirrors that of a charged particle in a Maxwell field, with the effective potential $\Phi_\mu^{(\mathrm{eff})}$ generating both scalar and vector gravitational interactions. No curved metric or covariant Laplacian is required: quantum dynamics proceeds directly from the same gauge principles that govern classical motion.

7.11.6 Significance

The quantum wave equation provides the foundation of the FSG test–particle sector. It connects seamlessly with the WKB analysis (whose leading term reproduces the Hamilton–Jacobi equation), and it supplies the building block for semiclassical quantisation conditions and for the study of bound gravitational states, wave packets, and scattering in subsequent subsections. In this way, particle quantisation in Flat Spacetime Gravity is every bit as natural and systematic as in electromagnetism, highlighting once again the consistency and versatility of the flat–space gauge formulation.

7.12 Concluding Remarks

The Hamiltonian formulation developed in this section confirms a striking and structurally important fact: a relativistic flat-spacetime gravitational field can be described in flat spacetime by a fully canonical field theory whose dynamics, constraints, and gauge structure mirror those of electromagnetism with mathematical fidelity. No curved metric, no Christoffel symbols, and no Riemann tensor are required in order to exhibit a consistent Hamiltonian evolution for a long–range flat-spacetime gravitational interaction.

The antisymmetric tensor $G_{\mu\nu}$ serves as the flat-spacetime gravitational analogue of the electromagnetic field tensor, and it carries with it all the advantages of a Maxwell–type theory: linear evolution, a simple constraint enforced by Φ_0, a clean decomposition into electric–like and magnetic–like components, and a Hamiltonian density quadratic in those fields. The canonical equations reproduce exactly the covariant system

$$\partial_\mu G^{\mu\nu} = -\frac{4\pi G}{c^2} J^\nu, \qquad \partial_{[\lambda} G_{\mu\nu]} = 0$$

with no modification, approximation, or geometric input.

The canonical structure revealed in this section is far more than a formal repackaging of earlier results. It shows that a flat-spacetime gravitational interaction can be cast as a flat–space gauge field with a well–defined Hamiltonian density, primary constraints, a Gauss law, and a covariant set of evolution equations. Nothing in the construction requires curvature, Christoffel symbols, or a dynamical metric: all flat-spacetime gravitational phenomena are encoded in the antisymmetric tensor $G_{\mu\nu}$ and the minimal coupling $\Phi_\mu J^\mu$. The Flat Spacetime Gravity therefore satisfies the strongest structural tests of a relativistic field theory: it is Lagrangian, Hamiltonian, constrained, gauge–invariant, and fully covariant. In this sense the theory achieves a remarkable synthesis: flat-spacetime gravitational dynamics emerges with the mathematical clarity of Maxwell's electrodynamics while remaining strictly rooted in Minkowski geometry. This canonical foundation not only ensures internal consistency but also opens the way for semiclassical, perturbative, and quantisation analyses entirely parallel to those used in the electromagnetic field.

The Hamiltonian structure achieved here reinforces and completes the picture developed in the previous sections: gravitation, when formulated through $G_{\mu\nu}$, becomes a clean, linear, gauge–theoretic field on flat geometry, endowed with the same canonical clarity as electromagnetism. This is a remarkable property—one that both reflects the internal coherence of the theory and opens the way for fully Hamiltonian analyses of flat-spacetime gravitational waves, radiation reaction, energy flux, stability, and quantisation in a Minkowskian setting.

In its final subsections, the Hamiltonian framework was shown to extend smoothly beyond purely classical dynamics. The Hamilton–Jacobi function admits a WKB interpretation whose rays coincide with the classical FSG trajectories, bounded orbits are described by action–angle variables that reproduce the perihelion advance, and the test particle can be quantized canonically via a Klein–Gordon–type equation in the effective potential. Thus, the Hamiltonian formulation not only passes all classical structural tests, but also dovetails naturally with the semiclassical and quantum developments of the subsequent chapters.

A relativistic field theory that possesses (i) a quadratic field Hamiltonian, (ii) canonical momenta, (iii) a Gauss constraint generated by a non–dynamical scalar potential, and (iv) a covariant Maxwell–type evolution system is, in principle, quantizable by the same canonical or path–integral techniques used for Maxwell theory. The flat-spacetime gravitational framework therefore satisfies all structural prerequisites for a flat–space quantum formulation.

8. De Broglie–Gravitation Connection

8.1 Introductory Remark

In the framework of Flat Spacetime Gravity the gravitational interaction is described entirely by the antisymmetric field tensor $G_{\mu\nu}$, living on a fixed Minkowski background. This structure has a remarkable and unique consequence: because the field is globally defined, antisymmetric, and governed by first–order Maxwell–type equations, one may legitimately integrate these equations over arbitrary 4–dimensional hypersurfaces. Such an operation has *no analogue* in General Relativity (where curvature replaces antisymmetric fields) and no analogue in Electromagnetism (where the source 7–current J^{μ} encodes charge flow rather matter 4–momentum P^{μ}).

The result is striking: the integrated Maxwell–Ampère equation of Flat Spacetime Gravity links the *total 4–momentum* of a compact source directly to a *hypersurface integral of the gravitational field tensor itself.* This identification feeds immediately into quantum theory via the de Broglie relation $P^{\mu} = \hbar K^{\mu}$, yielding an integrated-field representation of the wave 4–vector. The present section develops this connection carefully and systematically.

8.2 The Integrated Maxwell–Ampère Equation

The source equation of Flat Spacetime Gravity reads

$$\partial_{\mu} G^{\mu\nu} = -\frac{4\pi G}{c^2} J^{\nu}, \qquad J^{\nu} = \rho_0 U^{\nu}$$

We integrate both sides over a compact 4–volume $\mathcal{V}$ with boundary hypersurface $\partial\mathcal{V}$. Using Gauss's theorem in four dimensions,

$$\int_{\mathcal{V}} \partial_{\mu} G^{\mu\nu} d^4x = \int_{\partial\mathcal{V}} G^{\mu\nu} d\Sigma_{\mu\nu}$$

where $d\Sigma_{\mu\nu}$ is the oriented hypersurface element. Thus

$$\int_{\partial\mathcal{V}} G^{\mu\nu}\, d\Sigma_{\mu\nu} \;=\; -\,\frac{4\pi G}{c^2}\int_{\mathcal{V}} J^{\nu}\, d^4x$$

This identity is purely kinematic, relying only on the antisymmetry of $G_{\mu\nu}$ and on the existence of a fixed Lorentzian background. The right–hand side will shortly be identified with the total enclosed 4–momentum of the matter contained in $\mathcal{V}$, and the left–hand side therefore becomes an *integrated representation* of that momentum in terms of the gravitational field.

8.3 Hypersurface Elements and Physical Meaning of $\int G_{\mu\nu}\, d\Sigma^{\mu\nu}$

For any oriented three–dimensional hypersurface $\partial\mathcal{V}$ embedded in Minkowski spacetime, the surface element is

$$d\Sigma_{\mu\nu} \;=\; \frac{1}{2}\,\epsilon_{\mu\nu\alpha\beta}\, dx^{\alpha} \wedge dx^{\beta}$$

which is antisymmetric in (μ, ν). The contraction

$$G^{\mu\nu}\, d\Sigma_{\mu\nu}$$

is therefore a natural invariant flux of the gravitational field tensor through the hypersurface. In analogy with the electromagnetic flux $\int F_{\mu\nu} d\Sigma^{\mu\nu}$, but with crucial physical differences, the integral of $G_{\mu\nu}$ measures not merely a circulation-like quantity but the *gravitational content* of the matter flow crossing the hypersurface.

The identity of the previous subsection then provides

$$\int_{\partial\mathcal{V}} G^{\mu\nu}\, d\Sigma_{\mu\nu} \;=\; -\,\frac{4\pi G}{c^2}\int_{\mathcal{V}} J^{\nu}\, d^4x$$

indicating that the hypersurface integral of $G_{\mu\nu}$ encodes the entire 4–momentum budget of the enclosed matter. This is the starting point of the deeper result developed in the following sections: the emergence of the de Broglie invariants from an integrated classical gravitational field.

8.4 Identification of the Enclosed 4–Momentum

For a compact, spatially bounded source with rest–frame density ρ_0 and 4–velocity U^ν, the integrated 4–current is

$$\int_V J^\nu \, d^4x = \int_V \rho_0 \, U^\nu \, d^4x$$

If the integration spans exactly one proper-time layer of the matter worldtube, then the 4–volume integral of $\rho_0 U^\nu$ reduces to the total inertial 4–momentum:

$$P^\nu = m_0 U^\nu = \int_V \rho_0 \, U^\nu \, d^4x$$

Hence the source equation becomes

$$\int_{\partial V} G^{\mu\nu} \, d\Sigma_{\mu\nu} = -\frac{4\pi G}{c^2} \, P^\nu$$

The appearance of P^ν —and not merely the rest mass— is essential: it implies that information about both energy and momentum flows is encoded in the integrated flux of the FSG gravitational field. This already has no analogue in GR (where curvature, not an antisymmetric tensor, carries the field degrees of freedom), and no analogue in EM (where $\int F_{\mu\nu} d\Sigma^{\mu\nu}$ does *not* yield the matter 4–momentum).

8.5 Emergence of the de Broglie Relation

Quantum mechanics identifies the wave 4–vector K^ν of a particle with its 4–momentum via the de Broglie relation

$$P^\nu = \hbar K^\nu$$

Inserting this into the integrated source equation of Flat Spacetime Gravity yields:

$$\int_{\partial V} G^{\mu\nu}\, d\Sigma_{\mu\nu} \;=\; -\,\frac{4\pi G}{c^2}\;\hbar K^{\nu}$$

Thus, the wave-vector of a particle —the fundamental object governing quantum propagation, interference, and dispersion of matter waves— is encoded directly in the integrated gravitational field:

$$K^{\nu} \;=\; -\,\frac{c^2}{4\pi G\hbar}\int_{\partial V} G^{\mu\nu}\, d\Sigma_{\mu\nu}$$

This relation is conceptually extraordinary: a particle's quantum phase gradient is retrieved from a *classical* integral of the Flat Spacetime Gravity field tensor. No curvature, no metric expansion, no tetrads, and no quantum geometry are required. The wave–vector is a purely integral property of an antisymmetric field in flat spacetime.

8.6 The de Broglie Invariant as a Surface Integral of $G_{\mu\nu}$

The fundamental Lorentz scalar of quantum mechanics is

$$K_{\mu}K^{\mu} \;=\; -\,\frac{m_0^2 c^2}{\hbar^2}$$

Using the previous identity, one obtains

$$K_{\mu}K^{\mu} = \frac{c^4}{(4\pi G\hbar)^2}\left(\int_{\partial V} G_{\mu\nu}\, d\Sigma^{\mu\nu}\right)\left(\int_{\partial V} G^{\alpha\beta}\, d\Sigma_{\alpha\beta}\right)$$

Thus, the de Broglie invariant becomes:

$$K_{\mu}K^{\mu} \;=\; \frac{c^4}{(4\pi G\hbar)^2}\left\|\int_{\partial V} G_{\mu\nu}\, d\Sigma^{\mu\nu}\right\|^2$$

This is a purely classical integral representation of the quantum rest–mass invariant. Such a result is impossible in GR, where no antisymmetric gravitational field tensor exists, and where the wave 4–

vector of matter fields never arises as a geometric surface integral of curvature.

8.7 Quantum Phase as Integrated Gravitational Flux

The phase of a quantum particle is

$$\varphi = K_\mu X^\mu$$

and therefore, the integrated gravitational field over a hypersurface determines the phase accumulated by the particle as it crosses that hypersurface. Explicitly:

$$\varphi = -\frac{c^2}{4\pi G\hbar} X_\nu \int_{\partial V} G^{\mu\nu}\, d\Sigma_{\mu\nu}$$

Thus, the phase is literally a gravitational flux functional. Every interference pattern, every dispersion relation, every propagation phase of matter waves in a Flat Spacetime Gravity universe inherits structure from an integrated, classical, antisymmetric field. This offers an entirely novel classical geometric interpretation of de Broglie waves.

8.8 Why This Has No Analogue in GR or EM

- In GR, the fundamental field is the metric $g_{\mu\nu}$; curvature $R_{\mu\nu\alpha\beta}$ does not admit a global antisymmetric two–form that captures momentum flow through surface integrals.
- In EM, the antisymmetric field $F_{\mu\nu}$ *does* exist, but its source is J^ν, not the matter 4–momentum. Thus $\int F_{\mu\nu} d\Sigma^{\mu\nu}$ never yields P^ν.
- Only in Flat Spacetime Gravity do we have the coincidence:

$$J^\nu = \rho_0 U^\nu \quad\Rightarrow\quad \int J^\nu d^4x = P^\nu$$

making the integrated Maxwell–Ampère equation a direct bridge between classical field flux and quantum momentum.

8.9 Concluding Remarks

The chain of identities

$$\partial_\mu G^{\mu\nu} = -\frac{4\pi G}{c^2} J^\nu \Rightarrow \int_{\partial V} G^{\mu\nu}\, d\Sigma_{\mu\nu} = -\frac{4\pi G}{c^2} P^\nu$$
$$\Rightarrow \quad K^\nu = -\frac{c^2}{4\pi G\hbar}\int_{\partial V} G^{\mu\nu}\, d\Sigma_{\mu\nu}$$

reveals a structural feature that is unique in modern theoretical physics: *in flat–spacetime gravity, the classical gravitational field encodes the quantum wave–vector of matter as a hypersurface flux.*

This link is invisible in GR, absent in EM, and unavailable in any curvature-based formulation. It arises solely because Flat Spacetime Gravity employs an antisymmetric gravitational field with Maxwell–type dynamics. Few results illustrate so clearly the conceptual power of placing gravity in a Minkowski framework.

9. WKB Structure in FSG and Matter Waves

9.1 Introductory Remark

In the preceding sections the gravitational gauge field of the Flat Spacetime Gravity was developed within a fully classical Hamiltonian framework, culminating in a Hamilton–Jacobi formulation, a semiclassical WKB description, action--angle variables for bound orbits, and the canonical quantisation of the test particle sector. The present section extends this programme to the domain of quantum wave dynamics of the field itself and of its interaction with matter. Here we establish the semiclassical and quantum behavior of matter waves propagating in the effective gravitational gauge background.

Starting from the Bohr--Sommerfeld conditions and the WKB spectrum derived from the Hamilton--Jacobi structure, we develop the canonical field quantisation of the Flat Spacetime Gravity gauge potential, construct the corresponding interaction Hamiltonian, and analyze scattering, radiation emission, and gauge--invariant amplitudes. Ward identities, propagator structure, and radiative corrections are examined in close analogy with Abelian quantum field theory, demonstrating that the Flat Spacetime Gravity admits a fully consistent and predictive quantum framework without departure from Minkowski spacetime.

This section therefore constitutes the conceptual bridge between the semiclassical dynamics of individual trajectories and the quantum field description of gravitational interactions mediated by the Flat Spacetime Gravity gauge field.

9.2 Semiclassical Quantisation (Bohr–Sommerfeld Conditions)

9.2.1 Bohr–Sommerfeld quantisation

The action–angle formulation of the FSG Kepler problem, developed in the previous subsection, provides the natural setting for semiclassical quantisation. In the semiclassical (WKB) limit the quantum states of a bound system are associated with closed curves in phase space whose action integrals satisfy discrete quantisation conditions. For a central potential, the relevant actions are the radial and angular invariants (J_r, J_ϕ), and the Bohr–Sommerfeld prescription assigns them integer multiples of Planck's constant. This subsection applies that prescription to the effective FSG potential.

For a single periodic degree of freedom with action J, semiclassical quantisation imposes

$$J = nh, \qquad n \in \mathbb{Z}_{\geq 0}$$

For a planar bound orbit in a central potential, the independent classical actions are

$$J_\phi = \ell$$

$$J_r = \frac{1}{\pi} \int_{r_{\min}}^{r_{\max}} \sqrt{2m_0^2[E - \phi^{(\mathrm{eff})}(r)] - \frac{\ell^2}{r^2}}\; dr$$

Using $h = 2\pi\hbar$, the Bohr–Sommerfeld condition can equivalently be written as

$$J_\phi = \ell = \hbar m, \qquad m \in \mathbb{Z}$$

$$J_r = \hbar\,(n_r + \alpha), \qquad n_r \in \mathbb{Z}_{\geq 0}$$

where α is the Maslov index ($\alpha = 1/2$ for a radial oscillator with two turning points).

9.2.2 Effective potential in Flat Spacetime Gravity

The scalar potential entering the radial action is

$$\phi^{(\mathrm{eff})}(r) = -\frac{GM}{r} - \frac{GMh^2}{c^2 r^3}$$

consisting of the Newtonian part plus the orbital (relativistic) correction. Accordingly, the radial action separates into

$$J_r = J_r^{(0)} + \delta J_r$$

with $J_r^{(0)}$ the Newtonian action and δJ_r the FSG correction.

9.2.3 Energy levels and FSG corrections

Solving previous equations for the energy yields the semiclassical spectrum

$$E_{n_r,m} = E_{n_r,m}^{(0)} + \delta E_{n_r,m}$$

with the Newtonian term

$$E_{n_r,m}^{(0)} = -\frac{G^2M^2m_0^3}{2\hbar^2(n_r + m + \alpha)^2}$$

and the FSG correction generated by $\delta\phi(r)$:

$$\delta E_{n_r,m} = -\frac{3GMm_0h^2}{c^2(n_r + m + \alpha)^4}$$
$$\times \text{(numerical factor depending on the orbit)}$$

The exact form of δE depends on the detailed evaluation of δJ_r, but its qualitative structure is fixed: the orbital potential produces a downward shift of the spectrum, fully analogous to the relativistic ($\sim 1/c^2$) corrections in the Sommerfeld spectrum of the Hydrogen atom.

9.2.4 Connection with classical Flat Spacetime Gravity dynamics

The quantized orbits satisfy

$$\frac{\omega_\phi}{\omega_r} = 1 + O\left(\frac{GM}{c^2a(1-e^2)}\right)$$

and the splitting between the two frequencies reproduces the classical perihelion advance in the limit $n_r, m \gg 1$. Thus, the semiclassical spectrum encodes the same dynamical information obtained in the previous sections from the analysis of the orbital equation.

9.2.5 Interpretation

The Bohr–Sommerfeld framework reveals that the FSG corrections to Newtonian gravity manifest not only as classical perihelion shifts, but also as relativistic corrections to the semiclassical spectrum of bound states. This is a direct parallel with the transition from the Bohr model to the Sommerfeld fine structure in electromagnetism: the additional $1/r^2$ potential generated by the orbital term plays the role of a relativistic perturbation.

9.2.6 Significance

Semiclassical quantisation therefore integrates smoothly with the Hamiltonian and Hamilton–Jacobi formulations of the Flat Spacetime Gravity. The actions (J_r, J_ϕ), introduced at the classical level, become the quantised invariants of bound motion. This unifies the classical orbital dynamics with its quantum analogue and completes the particle sector of the FSG canonical structure, preparing the way for the quantisation of the *field* itself in the next subsection.

9.3 Canonical Quantisation of the Flat Spacetime Gravity Field

9.3.1 Canonical pair and primary constraint

The FSG interaction is mediated by a Maxwell–type antisymmetric tensor

$$G_{\mu\nu} = \partial_\mu \Phi_\nu - \partial_\nu \Phi_\mu$$

propagating on flat Minkowski spacetime. This makes the field sector structurally parallel to electromagnetism and allows a complete canonical quantisation based on the standard machinery of constraint Hamiltonian systems. In this subsection we develop the canonical quantisation of the *field* Φ_μ itself, independently of any test particle.

From the section of the Hamiltonian analysis, the canonical momentum conjugate to Φ_μ is

$$\Pi^\mu = \frac{\partial \mathcal{L}_{\text{field}}}{\partial(\partial_0 \Phi_\mu)} = \frac{c^2}{4\pi G} G^{0\mu}$$

The antisymmetry of $G_{\mu\nu}$ implies the primary constraint

$$\Pi^0 = 0$$

which is the analogue of the vanishing conjugate momentum of the temporal component of the electromagnetic potential A_0. This is the hallmark of a gauge theory: the scalar potential has no independent dynamical momentum and enforces the Gauss constraint.

9.3.2 Equal–time commutation relations

Canonical quantisation promotes (Φ_i, Π^i) to operators obeying the equal–time commutation relations

$$\left[\, \Phi_i(\mathbf{x}, t), \, \Pi^j(\mathbf{y}, t) \,\right] = i\hbar \, \delta_i^j \, \delta^3(\mathbf{x} - \mathbf{y})$$

with all other equal–time commutators vanishing. The temporal component Φ_0 acts as a Lagrange multiplier, and the constraint $\Pi^0 = 0$ is imposed as an operator equation.

9.3.3 Gauss constraint as a first–class generator

The field equations of this section show that the divergence

$$\mathcal{G}(\mathbf{x}) = \partial_i \Pi^i + \frac{4\pi G}{c^2} J^0(\mathbf{x})$$

must vanish on physical states:

$$\mathcal{G}(\mathbf{x})\,|\Psi_{\text{phys}}\rangle = 0$$

In the absence of sources ($J^\mu = 0$), the Gauss constraint simplifies to

$$\partial_i \Pi^i = 0$$

which is first class and generates gauge transformations via

$$\delta\Phi_\mu = \{\Phi_\mu, \int d^3x\, \Lambda \mathcal{G}\}$$

This confirms that the Flat Spacetime Gravity field carries the correct gauge freedom, precisely analogous to the electromagnetic field with gauge parameter Λ.

9.3.4 Hamiltonian and mode expansion

In the radiation gauge $\Phi_0 = 0$, $\nabla \cdot \mathbf{\Phi} = 0$ (which is fully admissible in flat spacetime), the Hamiltonian density reduces to

$$\mathcal{H} = \frac{2\pi G}{c^2}\,\Pi^i\Pi_i + \frac{c^2}{8\pi G}\,(\nabla \times \mathbf{\Phi})^2$$

One may expand the spatial potential in normal modes,

$$\Phi_i(\mathbf{x}, t) =$$

$$\sum_{\lambda=1}^{2} \int \frac{d^3k}{(2\pi)^{3/2}} \frac{1}{\sqrt{2\omega_{\mathbf{k}}}} \left[a_{\mathbf{k},\lambda}\, \epsilon_i(\mathbf{k},\lambda)\, e^{-i(\omega_{\mathbf{k}}t - \mathbf{k}\cdot\mathbf{x})} + a^\dagger_{\mathbf{k},\lambda}\, \epsilon_i^*(\mathbf{k},\lambda)\, e^{\,i(\omega_{\mathbf{k}}t - \mathbf{k}\cdot\mathbf{x})}\right]$$

with $\omega_{\mathbf{k}} = c|\mathbf{k}|$ and two transverse polarizations $\epsilon_i(\mathbf{k}, \lambda)$, satisfying

$$k^i \epsilon_i(\mathbf{k}, \lambda) = 0$$

The operators $a_{\mathbf{k},\lambda}$ and $a^\dagger_{\mathbf{k},\lambda}$ obey the standard commutation relations of a massless spin–1 field.

9.3.5 Physical interpretation

Quantisation of the Flat Spacetime Gravity field yields a massless, transverse spin–1 mediator of the gravitational interaction. Its excitations propagate at the speed of light, carry two independent polarizations, and obey the same field algebra as the photon, though with gravitational coupling strength G rather than electric charge e. The constraint structure ensures that only transverse degrees of freedom survive as physical states, a hallmark of gauge theories in flat spacetime.

9.3.6 Significance

Canonical field quantisation establishes that the Flat Spacetime Gravity is not merely a classical analogy to electromagnetism but a genuinely quantizable gauge field. Its canonical structure, constraints, commutators, and mode expansion all mirror those of Maxwell theory, demonstrating that gravitational phenomena can be described by a massless spin–1 field on Minkowski spacetime in a fully consistent quantum framework. This quantum foundation underlies the interaction Hamiltonian and scattering analysis developed in the next subsection.

9.4 Interaction Hamiltonian and Transition Amplitudes

9.4.1 Interaction Hamiltonian

The canonical quantisation of the Flat Spacetime Gravity field allows us now to construct the *interaction Hamiltonian* describing how a quantized flat-spacetime gravitational quantum couples to a test particle. The structure is the exact analogue of QED: the classical interaction term in the action,

$$S_{\text{int}} = m_0 \int \Phi_\mu^{(\text{eff})} U^\mu \, d\tau$$

after quantisation becomes an operator that mediates absorption and emission of field quanta. Because the effective potential separates into

$$\Phi_\mu^{(\mathrm{eff})} = \Phi_\mu + \Phi_\mu^{(\mathrm{orb})}$$

only the genuine field part Φ_μ participates in the quantum interaction; the orbital contribution is a classical c-number functional of the trajectory.

In canonical variables, the interaction Hamiltonian reads

$$H_{\mathrm{int}}(t) = -m_0\,\Phi_\mu(x(t))\,U^\mu(t)$$

with Φ_μ promoted to an operator field and $U^\mu(t)$ treated as the particle current. Writing the spatial part explicitly,

$$H_{\mathrm{int}}(t) = m_0\,\phi(\mathbf{x}(t),t)\,\gamma(t) - m_0\,\mathbf{A}(\mathbf{x}(t),t)\cdot\mathbf{u}(t)$$

reveals the exact homology with QED: a scalar coupling to the energy density, and a vector coupling to the three-velocity. This is the universal "minimal coupling" structure familiar from electromagnetism, now transplanted into the Flat Spacetime Gravity gauge theory.

9.4.2 Mode expansion and basic vertex

Substituting the mode expansion

$$\Phi_i(\mathbf{x},t) = \sum_{\lambda=1}^{2}\int \frac{d^3k}{(2\pi)^{3/2}}\frac{1}{\sqrt{2\omega_{\mathbf{k}}}}\Big[a_{\mathbf{k},\lambda}\,\epsilon_i(\mathbf{k},\lambda)\,e^{-i(\omega_{\mathbf{k}}t-\mathbf{k}\cdot\mathbf{x})} + a_{\mathbf{k},\lambda}^{\dagger}\,\epsilon_i^{*}(\mathbf{k},\lambda)\,e^{\,i(\omega_{\mathbf{k}}t-\mathbf{k}\cdot\mathbf{x})}\Big]$$

gives the elementary interaction vertex

$$H_{\mathrm{int}} \sim m_0\left[a_{\mathbf{k},\lambda}\,e^{-ik\cdot x(t)} + a_{\mathbf{k},\lambda}^{\dagger}\,e^{+ik\cdot x(t)}\right](\gamma\,\epsilon_0 - \mathbf{u}\cdot\boldsymbol{\epsilon})$$

where $k^\mu = (\omega_{\mathbf{k}}/c,\mathbf{k})$ and $\epsilon^\mu = (\epsilon_0,\boldsymbol{\epsilon})$ is the physical polarisation vector (transverse: $\mathbf{k}\cdot\boldsymbol{\epsilon} = 0$). The coefficient

$$\gamma\,\epsilon_0 - \mathbf{u}\cdot\boldsymbol{\epsilon}$$

is the gravitational analogue of the dipole coupling $j^\mu \epsilon_\mu$ in QED.

9.4.3 Transition amplitudes

The first-order transition amplitude between an initial state $|i\rangle = |\,\psi_i; n_{\mathbf{k},\lambda}\rangle$ and a final state $|f\rangle = |\,\psi_f; n_{\mathbf{k},\lambda} \pm 1\rangle$ is

$$\mathcal{A}_{i\to f} = -\frac{i}{\hbar}\int_{-\infty}^{+\infty} dt\,\langle f\,|\,H_{\text{int}}(t)\,|\,i\rangle$$

For single-quantum absorption, the relevant matrix element is

$$\langle\psi_f|\; m_0(\gamma(t)\,\epsilon_0^* - \mathbf{u}(t)\cdot\boldsymbol{\epsilon}^*)e^{ik\cdot x(t)}|\psi_i\rangle$$

while for emission one replaces $\epsilon^* \to \epsilon$ and $e^{ik\cdot x} \to e^{-ik\cdot x}$. Hence the transition probability depends on the Fourier transform of the classical current

$$j^\mu(t) = m_0\,U^\mu(t)$$

evaluated at the field's frequency $\omega_{\mathbf{k}}$.

9.4.4 Physical meaning

The Flat Spacetime Gravity predicts that gravitational radiation is emitted and absorbed by exactly the same kind of current that appears in the classical dynamical law. In the quantum description, the current $m_0 U^\mu$ acts as a source or sink of massless spin-1 quanta. The polarisation structure and selection rules coincide with those of a transverse gauge field, and all calculations proceed within flat spacetime without curved-metric corrections.

Thus, the interaction Hamiltonian elevates the Flat Spacetime Gravity field from a classical analogy to a fully quantum gauge interaction. It provides the foundation for cross-sections, decay rates, and scattering processes, exactly as in QED—with the gravitational constant G supplying the coupling strength.

9.5 Scattering, Cross–Sections, and the Born Approximation

9.5.1 S–matrix and Born approximation

With the interaction Hamiltonian of the previous subsection, the Flat Spacetime Gravity admits a fully quantum scattering framework. Because the field is linear, massless, transverse, and quantized by canonical commutators, the construction of transition amplitudes mirrors the QED formalism: the S–matrix is expanded in powers of the interaction and the leading term already captures the physics of weak gravitational scattering. All calculations proceed in flat Minkowski spacetime and rely only on the gauge–covariant coupling

$$H_{\text{int}}(t) \;=\; -\,m_0\,\Phi_\mu(x(t))\,U^\mu(t)$$

with Φ_μ the quantized FSG potential.

The scattering amplitude between asymptotic particle states $|\psi_i\rangle$ and $|\psi_f\rangle$ is defined by

$$S_{fi} \;=\; \delta_{fi} - \frac{i}{\hbar}\int dt\;\langle f\,|\,H_{\text{int}}(t)\,|\,i\rangle + O(H_{\text{int}}^2)$$

At weak coupling (small G), the first-Born approximation is accurate and the amplitude is dominated by the Fourier transform of the effective potential. Using the mode expansion of Φ_μ,

$$\Phi_\mu(x) = \sum_{\lambda=1}^{2}\int \frac{d^3k}{(2\pi)^{3/2}}\frac{1}{\sqrt{2\omega_{\mathbf{k}}}}\big[a_{\mathbf{k},\lambda}\,\epsilon_\mu(\mathbf{k},\lambda)e^{-ik\cdot x} + a_{\mathbf{k},\lambda}^\dagger\,\epsilon_\mu^*(\mathbf{k},\lambda)\,e^{+ik\cdot x}\big]$$

one obtains the basic vertex

$$\mathcal{M}(k,\lambda)\;\sim\;m_0\int dt\;U^\mu(t)\,\epsilon_\mu^*(\mathbf{k},\lambda)\,e^{ik\cdot x(t)}$$

This is the gravitational analogue of the QED dipole matrix element $j^\mu\epsilon_\mu^*$, with m_0U^μ playing the role of a mass current.

9.5.2 Static scattering and effective potential

For scattering from a static central source, one may use the classical effective potential

$$\phi^{(\text{eff})}(r) \;=\; -\frac{GM}{r} - \frac{GMh^2}{c^2 r^3}$$

which determines the long-range behavior of the interaction. Its Fourier transform,

$$\tilde{\phi}^{(\text{eff})}(\mathbf{q}) \;=\; \int d^3x\; e^{-i\mathbf{q}\cdot\mathbf{x}}\, \phi^{(\text{eff})}(r)$$

$$\mathbf{q} = \mathbf{k}_f - \mathbf{k}_i$$

immediately yields the Born-level scattering amplitude

$$\mathcal{M}_{fi} \;=\; -m_0\, \tilde{\phi}^{(\text{eff})}(\mathbf{q})$$

with the corresponding differential cross–section

$$\frac{d\sigma}{d\Omega} \;=\; \frac{|\mathcal{M}_{fi}|^2}{(4\pi\hbar v_\infty)^2}, \qquad v_\infty = \text{asymptotic velocity}$$

The Coulombic piece produces the usual Rutherford-type tail, while the $1/r^2$ orbital term modifies the forward-angle behaviour in a distinctive and observationally testable manner.

9.5.3 Emission and absorption of quanta

The same formalism describes transitions

$$|n_{\mathbf{k},\lambda}\rangle \;\leftrightarrow\; |n_{\mathbf{k},\lambda} \pm 1\rangle$$

with amplitudes governed by the Fourier components of the current $m_0 U^\mu(t)$. Periodic or quasi-periodic motion produces discrete emission lines, with selection rules controlled by the transversality of the polarization vectors and by the geometry of the orbit. In particular, circular orbits yield dominant dipole-type radiation, while eccentric orbits generate higher harmonics in analogy with QED bremsstrahlung.

9.5.4 Significance

The Born-level scattering analysis confirms that the Flat Spacetime Gravity field behaves as a massless spin–1 mediator with long-range interactions. Its Fourier structure reproduces the classical potentials of the previous sections, while the quantized description predicts definite amplitudes for radiation, absorption, and scattering—all within flat spacetime and without curved-metric modifications. This establishes a fully quantum–consistent, gauge–invariant foundation for the next subsection, where the analogy with QED becomes even sharper through Ward identities and gauge invariance.

9.6 Gauge Invariance and Ward Identities

9.6.1 Gauge transformations and the physical Hilbert space

A decisive test of any gauge field theory is whether its quantized interaction amplitudes satisfy the analogue of the *Ward identities*. These identities guarantee that unphysical polarizations do not contribute to observable processes, that scattering amplitudes depend only on the transverse degrees of freedom of the field, and that the coupling to matter is consistent with gauge invariance at every order. In the Flat Spacetime Gravity framework all these properties hold exactly, because the theory is built entirely on the gauge–covariant combination

$$p_\mu \; - \; m_0 \Phi_\mu$$

and on the antisymmetric field tensor

$$G_{\mu\nu} \;\; = \;\; \partial_\mu \Phi_\nu - \partial_\nu \Phi_\mu$$

A gauge transformation is defined by

$$\Phi_\mu \;\; \rightarrow \;\; \Phi_{\mu'} = \Phi_\mu + \partial_\mu \Lambda$$

with $\Lambda(x)$ an arbitrary scalar field. At the operator level this transformation is generated by the Gauss constraint

$$\mathcal{G}(\mathbf{x}) \;=\; \partial_i \Pi^i + \frac{4\pi G}{c^2} J^0(\mathbf{x})$$

acting on the Hilbert space as

$$|\Psi\rangle \;\to\; \exp\left[\frac{i}{\hbar}\int d^3x\, \Lambda(\mathbf{x})\, \mathcal{G}(\mathbf{x})\right]|\Psi\rangle$$

Physical states obey

$$\mathcal{G}(\mathbf{x})\,\big|\Psi_{\text{phys}}\big\rangle \;=\; 0$$

ensuring that longitudinal modes do not appear in any physical amplitude.

9.6.2 Vertex structure and gauge invariance

The interaction Hamiltonian,

$$H_{\text{int}} \;=\; -\,m_0\,\Phi_\mu(x)\,U^\mu(x)$$

is gauge invariant because under $\Phi_\mu \to \Phi_\mu + \partial_\mu\Lambda$ one has

$$\begin{aligned} \delta H_{\text{int}} \;&=\; -\,m_0\,\big(\partial_\mu\Lambda\big)\,U^\mu \;=\; -\,m_0\,\frac{d\Lambda}{d\tau} \\ \Rightarrow\quad \int dt\,\delta H_{\text{int}} &= -m_0\,[\Lambda(t)]_{t=-\infty}^{t=\infty} \;=\; 0 \end{aligned}$$

provided Λ vanishes at temporal infinity. Thus, the vertex factor entering scattering amplitudes is invariant under gauge transformations.

9.6.3 Ward identities for external legs

For an on–shell Flat Spacetime Gravity quantum with 4–momentum $k^\mu = (\omega_{\mathbf{k}}/c, \mathbf{k})$ and polarization $\epsilon^\mu(\mathbf{k},\lambda)$, the transversality condition

$$k_\mu\,\epsilon^\mu(\mathbf{k},\lambda) \;=\; 0$$

must imply that replacing ϵ^μ by k^μ annihilates the matrix element. Indeed, for any transition amplitude,

$$\mathcal{M}_{fi} = \int dt \; m_0 \, U^\mu(t) \, \epsilon^*_\mu(\mathbf{k}, \lambda) \, e^{ik\cdot x(t)}$$

one finds

$$\begin{aligned}\mathcal{M}_{fi}(\epsilon^\mu \to k^\mu) &= m_0 \int dt \; U^\mu(t) \, k_\mu \, e^{ik\cdot x(t)} \\ &= m_0 \int dt \; \frac{d}{dt}\left(e^{ik\cdot x(t)}\right) = 0\end{aligned}$$

since the exponential factor vanishes at temporal infinity. This is the FSG analogue of the Ward identity familiar from QED.

9.6.4 Internal Ward identities

At higher orders in perturbation theory, one encounters loop diagrams and vertex corrections. Because every interaction is mediated by the gauge–covariant current $m_0 U^\mu$ and because the propagator of the Φ_μ field is transverse in radiation gauge,

$$D_{\mu\nu}(k) = \frac{1}{k^2 + i0^+}\left(\eta_{\mu\nu} - \frac{k_\mu k_\nu}{|\mathbf{k}|^2}\right)$$

each internal line is automatically projected onto the subspace orthogonal to k^μ. Consequently, loop integrals inherit the identity

$$k^\mu \Gamma_\mu(k, p) = 0$$

for every proper vertex function Γ_μ, ensuring conservation of the gravitational mass current and protecting the renormalization of the coupling.

9.6.5 Significance

Gauge invariance pervades the entire quantum structure of the Flat Spacetime Gravity. The external Ward identities guarantee that only the transverse physical polarizations propagate to asymptotic states; the internal identities ensure consistency of the perturbation series and protect the theory from unphysical divergences. Thus, the quantized Flat Spacetime Gravity field satisfies *all* consistency conditions normally associated with canonical gauge theories, but within the geometric simplicity of flat spacetime.

The next subsection exploits these properties to construct a fully covariant propagator and to analyze radiative corrections in the Flat Spacetime Gravity setting.

9.7 Quantized Flat Spacetime Gravity Field

9.7.1 Field equation and gauge fixing

The quantisation of the Flat Spacetime Gravity gauge field requires the construction of a Green function (propagator) for the massless potential Φ_μ. Because the theory is Abelian and defined on flat Minkowski spacetime, the propagator shares the structural simplicity of the electromagnetic one. In this subsection we formulate both the covariant and the radiation–gauge propagators and show how the gauge structure enforces transversality.

Varying the action with respect to Φ_μ gives the source equation

$$\partial_\mu G^{\mu\nu} = -\frac{4\pi G}{c^2} J^\nu$$

To obtain an invertible kinetic operator one introduces a gauge–fixing condition. In covariant gauge ($\partial_\mu \Phi^\mu = 0$) the equation becomes

$$\Phi^\nu = -\frac{4\pi G}{c^2} J^\nu, \qquad \Box = \partial_\mu \partial^\mu$$

9.7.2 Covariant Feynman propagator

The propagator $D_{\mu\nu}(x - x')$ is defined by

$$D_{\mu\nu}(x - x') = \eta_{\mu\nu}\, \delta^4(x - x')$$

Its Fourier transform is

$$D_{\mu\nu}(k) = \frac{-\eta_{\mu\nu}}{k^2 + i0^+}$$

where $k^2 = \eta_{\alpha\beta} k^\alpha k^\beta$. This is the exact analogue of the Feynman propagator of QED, with the substitution $e \to \sqrt{G}$ in the coupling strength.

9.7.3 Radiation–gauge propagator

For physical processes involving asymptotic quanta, only transverse modes propagate. Imposing

$$\Phi_0 = 0, \qquad \nabla \cdot \boldsymbol{\Phi} = 0$$

one obtains the spatial propagator

$$D_{ij}(\boldsymbol{k}, \omega) = \frac{1}{\omega^2/c^2 - |\boldsymbol{k}|^2 + i0^+}\left(\delta_{ij} - \frac{k_i k_j}{|\boldsymbol{k}|^2}\right)$$

which projects automatically onto the two physical polarizations of the massless spin–1 field.

9.7.4 Gauge structure and consistency

The propagator satisfies

$$k^\mu D_{\mu\nu}(k) = 0, \qquad D_{\mu\nu}(k) k^\nu = 0$$

ensuring that unphysical longitudinal components decouple from all scattering amplitudes. These identities, together with the Ward relations of the previous subsection, guarantee the internal consistency of the quantized Flat Spacetime Gravity gauge field and validate the use of the propagator in loop corrections and radiative processes.

9.7.5 Significance

The existence of a covariant propagator firmly establishes the Flat Spacetime Gravity field as a fully quantizable massless gauge field. Propagators allow the construction of perturbation theory, renormalization of composite operators, and the calculation of radiative corrections—all within the flat Minkowski framework.

9.8 Effective Flat Spacetime Gravity Coupling

9.8.1 Vacuum polarization

With the covariant propagator in place, one may examine the structure of radiative corrections in the quantized Flat Spacetime Gravity gauge field. Because the theory is Abelian, massless, and formulated in flat Minkowski spacetime, its quantum loop structure is strikingly similar to that of QED. In this subsection we highlight the essential features: the modification of the effective coupling, the transverse structure of the vacuum polarization tensor, and the finiteness of physical amplitudes ensured by the Ward identities.

The leading radiative correction to the FSG propagator arises from the vacuum polarization tensor $\Pi_{\mu\nu}(k)$, defined via the dressed propagator

$$D^{(\mathrm{eff})}_{\mu\nu}(k) \;=\; D_{\mu\nu}(k) + D_{\mu\alpha}(k)\,\Pi^{\alpha\beta}(k)\,D_{\beta\nu}(k) + \cdots.$$

Gauge invariance imposes the transversality condition

$$k^{\mu}\Pi_{\mu\nu}(k) = 0, \qquad \Pi_{\mu\nu}(k)\,k^{\nu} = 0$$

which in turn implies the general decomposition

$$\Pi_{\mu\nu}(k) \;=\; \left(\eta_{\mu\nu} - \frac{k_\mu k_\nu}{k^2}\right)\Pi(k^2)$$

Thus, all loop corrections preserve the transverse projector of the free theory and do not generate a mass term for the gauge field.

9.8.2 One–loop correction and running coupling

At one loop, particle–field fluctuations generate an effective momentum–dependent coupling,

$$G_{\text{eff}}(k^2) = \frac{G}{1 - \Pi(k^2)}$$

the analogue of the running electromagnetic coupling $e_{\text{eff}}(q^2)$ in QED. The detailed form of $\Pi(k^2)$ depends on the mass spectrum of the FSG–coupled particles, but its qualitative effect is universal: the interaction strength between currents is modified according to the scale set by the momentum transfer.

Because the Flat Spacetime Gravity is Abelian and the coupling has dimensions of $(\text{length})^2$, loop corrections do not lead to Landau–pole–type pathologies in the physical scattering domain. The renormalized coupling is finite at all finite k^2.

9.8.3 Vertex corrections

The correction to the particle–field vertex takes the form

$$\Gamma_\mu(p', p) = U_\mu + \delta\Gamma_\mu(p', p)$$

with the Ward identity enforcing

$$k^\mu \Gamma_\mu(p', p) = 0$$

Thus, the longitudinal part of the vertex correction must vanish and $\delta\Gamma_\mu$ is necessarily transverse. This protects the renormalization of the effective coupling G_{eff} and ensures gauge–consistent scattering amplitudes.

9.8.4 Self–energy corrections

The self–energy of a FSG–coupled particle modifies its dispersion relation,

$$p_\mu p^\mu = -m_0^2 c^2 + \Sigma(p^2)$$

but because the interaction is linear in the current U^μ rather than quadratic in the field derivatives, these corrections do not spoil the structure of the Hamilton–Jacobi equation nor the definition of the canonical momentum. They produce finite mass renormalization analogous to the QED electron self–energy.

9.8.5 Physical significance

Radiative corrections in the Flat Spacetime Gravity exhibit the characteristic features of any well–behaved Abelian gauge field:

- the vacuum polarization is transverse,
- the propagator retains its projector structure,
- vertex corrections satisfy the Ward identity,
- the effective coupling $G_{\text{eff}}(k^2)$ is finite at all finite scales,
- physical amplitudes depend only on the two transverse polarizations.

All of this takes place without curved geometry, and without invoking the metric as a dynamical object. The quantized Flat Spacetime Gravity field thus admits a consistent perturbative expansion in flat spacetime, paralleling QED in its structure while differing only in the nature of the coupling constant and the interpretation of the current.

9.9 Semiclassical Radiation: Power, Spectra, and Selection Rules

9.9.1 Transition amplitude and power

With the quantum interaction Hamiltonian and the propagator machinery in place, one can now extract semiclassical predictions for gravitational radiation emitted by moving test particles in the Flat Spacetime Gravity framework. The structure is again parallel to QED: the current $m_0 U^\mu(t)$ acts as the classical source, while the quantized Flat Spacetime Gravity field determines the spectral distribution, angular pattern, and selection rules of the emitted quanta. All calculations occur in flat Minkowski spacetime, with no curved–metric corrections required.

For a particle following a worldline $x^\mu(t)$, the amplitude for emission of a Flat Spacetime Gravity quantum with momentum k^μ and polarization $\epsilon^\mu(\mathbf{k}, \lambda)$ is

$$\mathcal{A}_{\mathbf{k},\lambda} = -\frac{i m_0}{\hbar} \int_{-\infty}^{+\infty} dt \; U^\mu(t) \, \epsilon_\mu^*(\mathbf{k}, \lambda) \, e^{ik\cdot x(t)}$$

The corresponding differential probability per unit time is

$$dW_{\mathbf{k},\lambda} = \frac{1}{T} \, |\mathcal{A}_{\mathbf{k},\lambda}|^2 \, \frac{d^3k}{(2\pi)^3}$$

and the radiated power follows from

$$dP = \hbar\omega_{\mathbf{k}} \, dW_{\mathbf{k},\lambda}$$

Thus, the power spectrum is governed by the Fourier transform of the classical current, exactly as in electromagnetic bremsstrahlung.

9.9.2 Harmonic structure for periodic motion

If the motion is periodic with frequency Ω (e.g. bound orbits, circular or quasiperiodic trajectories), then

$$U^{\mu}(t) = \sum_{n=-\infty}^{+\infty} U_n^{\mu} \; e^{in\Omega t}$$

and the amplitude becomes

$$\mathcal{A}_{\mathbf{k},\lambda} = -\frac{im_0}{\hbar}\sum_n U_n^{\mu} \; \epsilon_{\mu}^{*}(\mathbf{k},\lambda) \int dt \; e^{i(\omega_{\mathbf{k}}-n\Omega)t}$$

Hence emission occurs at discrete harmonics

$$\omega_{\mathbf{k}} = n\Omega, \qquad n = 1,2,3,\ldots$$

with strengths proportional to $|U_n^{\mu}\epsilon_{\mu}^{*}|^2$. Circular orbits produce a dominant $n = 1$ line (dipole–type), while eccentric orbits generate higher harmonics.

9.9.3 Angular pattern and polarization

Because the Flat Spacetime Gravity field is massless and transverse, only the spatial components of ϵ^{μ} contribute. In radiation gauge the amplitude reduces to

$$\mathcal{A}_{\mathbf{k},\lambda} \propto \int dt \; \mathbf{u}(t) \cdot \boldsymbol{\epsilon}^{*}(\mathbf{k},\lambda) \, e^{ik\cdot x(t)}$$

Thus:

- No radiation is emitted along the instantaneous direction of motion.
- The pattern is purely transverse: $\mathbf{k} \cdot \boldsymbol{\epsilon} = 0$.
- Polarization dependence reflects the geometry of the orbit, as in electromagnetic dipole emission.

9.9.4 Selection rules

Gauge invariance and transversality enforce the analogue of electromagnetic selection rules:

- Only transitions that change the spatial part of the current $m_0\mathbf{u}(t)$ couple to radiation.
- Radiation vanishes for strictly radial (1D) motion.
- For symmetric orbits, only harmonics compatible with the rotational symmetry appear.
- The lowest nonzero multipole is dipole, not quadrupole, because the mediator is a spin–1 gauge field.

9.9.5 Power spectrum and flat–space consistency

The total radiated power can be written in compact form as

$$P = \frac{m_0^2 G}{\pi c^2} \sum_{\lambda=1}^{2} \int d^3k \, |\mathbf{u}_\perp(\omega_\mathbf{k})|^2 \delta(\omega_\mathbf{k} - n\Omega)$$

where $\mathbf{u}_\perp$ is the component of the velocity orthogonal to $\mathbf{k}$. The flat–space nature of the theory is manifest: all radiation characteristics follow directly from the current $U^\mu(t)$ and the transverse projector of the propagator, without a curved metric or geodesic deviation.

9.9.6 Significance

Semiclassical radiation in the Flat Spacetime Gravity is the exact analogue of QED bremsstrahlung: discrete harmonic emission for periodic trajectories, transverse polarization selection rules, and a power spectrum controlled by the Fourier components of the particle current. This provides a clean and experimentally testable prediction in scenarios where gravitational dipole–type radiation might arise—yet obtained entirely within flat Minkowski spacetime.

9.10 Concluding Remarks

In this section the semiclassical and quantum wave structure of the Flat Spacetime Gravity has been developed in a systematic and self-consistent manner. Beginning with the Bohr--Sommerfeld quantisation of bound orbits and the WKB spectrum derived from the Hamilton--Jacobi formulation, the formalism was extended to the canonical quantisation of the gravitational gauge field and the construction of interaction Hamiltonians for matter--field coupling. Scattering processes, radiation emission, and transition amplitudes were analyzed using standard operator techniques, while gauge invariance was ensured through the Gauss constraint and the associated Ward identities. The existence of a covariant propagator and the structure of radiative corrections demonstrated that the Flat Spacetime Gravity gauge field behaves as a fully quantizable massless spin--1 field defined on flat spacetime, free of inconsistencies and without the introduction of a dynamical spacetime metric.

The results of this section provide the complete semiclassical and quantum foundation for the theory. They establish a continuous line of development from classical Hamiltonian dynamics to quantum wave processes and radiation phenomena, thereby preparing the ground for the fully quantum canonical structure of the Flat Spacetime Gravity to be addressed in the following section.

10. Quantum Structure of Flat Space Gravity

10.1 Introductory Remark

The preceding section established the semiclassical wave dynamics of matter and gravitational radiation within the WKB framework of the Flat Space Gravity. In the present section we proceed to the fully quantum description of gravitational interactions by developing the operator formulation of bound states, light deflection, orbital stability, and gravitational–wave emission. Here the classical and semiclassical structures derived earlier are promoted to genuine quantum observables. Discrete spectra of bound orbits are obtained through canonical quantisation, quantum corrections to perihelion advance and light bending are evaluated, and the stability of circular orbits is analyzed at the operator level. Finally, the Flat Space Gravity gauge field itself is quantized, and its normal modes are interpreted as gravitational wave quanta carrying helicity.

This section therefore completes the transition from the WKB matter–wave description to the full quantum field–theoretic structure of the Flat Space Gravity, establishing a consistent quantum account of gravitational dynamics entirely within flat Minkowski spacetime.

10.2 Bound–State Spectra and the WKB–Bohr Correspondence

10.2.1 Effective radial potential and turning points

The semiclassical machinery developed in the previous subsections allows one to extract discrete spectral information from bound orbits in the Flat Space Gravity framework. As in electromagnetism, the key principle is that periodic or quasiperiodic motion generates a discrete harmonic structure in the classical current; in quantum theory these harmonics correspond to allowed transition frequencies between energy levels. In this subsection we show how the WKB quantization condition, the structure of the effective potential $\phi^{(\mathrm{eff})}$, and the harmonic content of the motion combine to yield a fully consistent semiclassical spectrum.

For a test particle in a central Flat Space Gravity field, the classical dynamics reduces to the effective radial equation

$$\frac{1}{2}m_0\left(\frac{dr}{dt}\right)^2 + V_{\text{eff}}(r) \;=\; E_{\text{orb}}$$

where

$$V_{\text{eff}}(r) = m_0\phi^{(\text{eff})}(r) = -\frac{GMm_0}{r} - \frac{GMm_0h^2}{c^2r^3}$$

For bound trajectories, the motion oscillates between two classical turning points r_{min} and r_{max} determined by

$$E_{\text{orb}} \;=\; V_{\text{eff}}(r)$$

10.2.2 WKB quantization

The semiclassical quantization of the radial motion follows the standard Bohr–Sommerfeld rule,

$$\oint p_r(r)\,dr \;=\; 2\pi\hbar(n_r + 1/2), \qquad n_r = 0,1,2,\ldots$$

where the radial momentum is

$$p_r(r) \;=\; m_0\gamma(r)\,\frac{dr}{dt}$$

and the integral is taken between the turning points. Because the effective potential contains both a $1/r$ and a $1/r^2$ term, the WKB spectrum differs from the Kepler spectrum of Newtonian gravity in a characteristic manner. The additional $1/r^2$ contribution shifts all levels in a way analogous to the fine structure of the hydrogen atom created by spin–orbit coupling.

10.2.3 Angular quantization

The angular momentum $h = |\mathbf{r} \times \mathbf{u}|$ is conserved in the FSG dynamics. Thus, the total semiclassical quantum number is

$$n = n_r + \ell + 1, \qquad \ell = 0,1,2,\dots$$

and the WKB quantization yields a discrete set of energies

$$E_{n\ell} = E_{\rm orb}(n,\ell) = -\frac{1}{2}\frac{(GMm_0)^2}{\hbar^2}\frac{1}{n^2}\left[1 + \alpha\,\frac{\ell(\ell+1)}{n^2}\right]$$

where α is a dimensionless coefficient encoding the strength of the $1/r^2$ correction

$$\alpha = \frac{h^2}{c^2\hbar^2}$$

The term proportional to α represents the FSG "fine structure" of the spectrum, directly traceable to the orbital piece of the effective potential.

10.2.4 Bohr correspondence and emission frequencies

The transition frequency for a semiclassical transition $(n,\ell) \to (n',\ell')$ is

$$\omega_{n\ell\to n'\ell'} = \frac{1}{\hbar}(E_{n\ell} - E_{n'\ell'})$$

These quantum transition frequencies coincide with the harmonics of the classical current:

$$\omega_{n\ell\to n'\ell'} = m\Omega_{\rm orb}, \qquad m \in \mathbb{Z}$$

where $\Omega_{\rm orb}$ is the fundamental orbital frequency and m labels the classical Fourier components of $U^\mu(t)$. This agreement confirms the full semiclassical correspondence between the WKB spectrum and the radiation spectrum of this section.

10.2.5 Polarization selection rules

The quantized Flat Space Gravity field couples to the transverse spatial components of the current. Thus, transitions must satisfy the dipole-type rule

$$\Delta\ell = \pm 1, \qquad m_{\ell'} = m_\ell \pm 1$$

with no monopole radiation allowed. These selection rules follow from the vector nature of the FSG mediator and are identical in structure to the electromagnetic dipole rules.

10.2.6 Significance

The WKB/Bohr analysis reveals that bound states in the Flat Space Gravity possess a discrete spectrum modified by the orbital component of the effective potential. The agreement between the classical harmonic content of the motion and the quantum transition frequencies provides a clean semiclassical validation of the quantized Flat Space Gravity field. Unlike general-relativistic treatments—where bound-state spectra require curved metrics or geodesic deviations—the entire analysis here is carried out in flat Minkowski space, with quantization arising solely from the gauge structure of Φ_μ.

10.3 Quantum Corrections to the Classical Perihelion Shift

10.3.1 WKB formulation

The FSG perihelion advance arises, as we have seen, from the *effective* potential

$$\phi^{(\text{eff})}(r) = -\frac{GM}{r} - \frac{GMh^2}{c^2r^3}$$

where the second term is the orbital ("spin–orbit–like'') correction. This term is entirely responsible for the classical precession.

Once the theory is quantized, the orbital angular momentum becomes discrete,

$$h \rightarrow \hbar\sqrt{\ell(\ell+1)}, \qquad \ell = 0,1,2,\ldots$$

and this replacement induces a *quantum correction* to the precession.

The radial motion in the effective potential is governed by the WKB quantization condition

$$\oint p_r(r)\,dr = 2\pi\hbar(n_r + 1/2)$$

with

$$p_r^2(r) = 2m_0\left[E - \left(-\frac{GM}{r} - \frac{GMh^2}{c^2r^3}\right) - \frac{h^2}{2m_0r^2}\right]$$

Expanding the effective radial frequency around the classical orbit gives

$$\Omega_{\text{aps}}(\ell) = \Omega_{\text{orb}}\left[1 + \frac{3GM}{a(1-e^2)c^2} + \beta\,\frac{\hbar^2}{\ell(\ell+1)\,m_0^2\,c^2\,a^4}\right]$$

where the last term originates *entirely* from the quantum replacement $h^2 \mapsto \hbar^2\ell(\ell+1)$ *inside the* $GMh^2/(c^2r^3)$ *term*. The factor a^{-4} (rather than a^{-2}) reflects precisely the $1/r^3$ structure of $\phi^{(\text{eff})}$.

10.3.2 Quantum correction to the shift

The total apsidal advance per orbit becomes

$$\Delta\varphi = \Delta\varphi_{\text{cl}} + \Delta\varphi_{\text{q}}$$

with

$$\Delta\varphi_{\text{cl}} = \frac{6\pi GM}{a(1-e^2)c^2}$$

and

$$\Delta\varphi_{\text{q}} = 2\pi\,\beta\,\frac{\hbar^2}{\ell(\ell+1)\,m_0^2\,c^2\,a^4}$$

The a^{-4} suppression shows that the quantum contribution is even more heavily suppressed than one might guess from a naive $1/r^2$ potential: the extra $1/r$ makes the correction scale like the fourth inverse power of the orbital radius.

10.3.3 Physical size of the effect

For planetary orbits, the quantum term is utterly negligible:

$$\frac{\Delta\varphi_{\mathrm{q}}}{\Delta\varphi_{\mathrm{cl}}} \sim \frac{\hbar^2}{m_0^2 c^2 a^4} \ll 10^{-70} \quad \text{(Mercury)}$$

This suppression is fully expected: macroscopic orbital scales exceed the particle's Compton wavelength by some 10^{40}–10^{50} orders of magnitude.

10.3.4 Conceptual significance

Even though the effect is unmeasurably small, its *existence* is conceptually important:

- it arises automatically from the quantized FSG dynamics, - it depends solely on the orbital potential,
- it confirms internal consistency between the *classical* and *quantum* versions of the theory, - it directly parallels the familiar quantum corrections to orbital motion in QED (e.g. spin–orbit and Lamb–type corrections), but here in flat spacetime and without curved geometry.

Thus, the perihelion shift—classical or quantum—emerges naturally from the flat–space Flat Space Gravity field, with quantum discreteness adding only a tiny but fully well–defined correction.

10.4 Quantum Light–Deflection in the Flat Space Gravity Field

10.4.1 Field–theoretic picture

In previous sections the deflection of light was obtained entirely within the classical FSG dynamics, using the effective potential

$$\phi^{(\text{eff})}(r) = -\frac{GM}{r} - \frac{GMh^2}{c^2r^3}$$

The first term produces the familiar Newtonian $1/bu^2$ contribution, which as we saw vanishes for photons, while the second ("orbital'') term produces the universal, velocity–independent bending equal to the same value as in General Relativity.

Here we examine the *quantum corrections* to this phenomenon within the quantized Flat Space Gravity.

In the quantum description, the bending of a photon corresponds to tree–level scattering in the external Flat Space Gravity field of a massive source. The amplitude is governed exclusively by the two transverse polarizations of the quantized field:

$$\mathcal{M} \propto \epsilon_{\mu}^{(\lambda)}(k)\ G^{\mu\nu}(q)\ \epsilon_{\nu}^{(\lambda')}(k')$$

with $q = k' - k$ the momentum transfer. Because the propagator remains transverse,

$$k^{\mu}D_{\mu\nu}(k) = 0$$

no longitudinal or gauge–dependent contributions appear.

10.4.2 Semiclassical extraction of the bending

Using the Born approximation and the potential $V(r) = m_0\phi^{(\text{eff})}(r)$, the leading scattering angle is

$$\theta(b) = -\frac{1}{p}\int_{-\infty}^{\infty}\frac{\partial V}{\partial r}\frac{b\,dz}{(b^2+z^2)^{\frac{1}{2}}}$$

$$r = \sqrt{b^2+z^2}$$

For a photon ($p = \hbar\omega/c$), the Newtonian term contributes as

$$\theta_{\text{N}} = \frac{2GM}{bv^2}$$

which *vanishes identically* for light. This remarkable cancellation is forced by the internal consistency of the special–relativistic limit: the "$v \to c$" limit corresponds to an "infinite denominator" in the Newtonian term, forcing it to zero.

The orbital part produces

$$\theta_{\text{orb}} = \frac{4GM}{bc^2}$$

independent of the photon energy. This is the *entire* classical bending for light in the Flat Space Gravity.

10.4.3 Quantum corrections

Quantum fluctuations (vacuum polarization, vertex corrections, and self–energy effects) modify the amplitude by replacing G with the momentum–dependent effective coupling

$$G_{\text{eff}}(q^2) = \frac{G}{1 - \Pi(q^2)}$$

where $\Pi(q^2)$ is the scalar vacuum–polarization function. Thus, the quantum bending angle becomes

$$\theta_{\text{q}} = \frac{4\, G_{\text{eff}}(q^2)\, M}{b\, c^2}$$

with $q \sim \hbar/b$ the typical momentum transfer. Since for astronomical impact parameters b is macroscopic, q is extremely small and

$$G_{\text{eff}}(q^2) = G\,[\,1 + \mathcal{O}(10^{-60})\,]$$

so, the correction is *utterly negligible*:

$$\frac{\theta_{\text{q}} - \theta_{\text{cl}}}{\theta_{\text{cl}}} \sim 10^{-60} \qquad \text{or smaller}$$

10.4.4 Interpretation

Even though the corrections are unobservable, their presence confirms:

- the Flat Space Gravity quantum field is internally consistent,
- radiative corrections deform the effective coupling in the expected QED–like way,
- the *absence* of a Newtonian contribution for photons remains exact at the quantum level,
- the velocity–independent $1/r^3$ orbital potential fully controls the deflection in both classical and quantum regimes.

10.4.5 Significance

Light bending in the quantized Flat Space Gravity is thus a rare case where the *classical prediction is already final to machine precision* at any realistic physical scale. Quantum corrections exist, are well–defined, and follow the flat–space Abelian gauge structure of the theory, but they are suppressed so strongly that the classical FSG prediction remains exact for all practical purposes.

10.5 Quantum Stability of Circular Orbits

10.5.1 Quantum effective potential

The last quantum effect we examine concerns the *stability* of circular orbits in the Flat Space Gravity field when the *orbital angular momentum is quantized.* Classically, circular orbits correspond to extrema of the effective potential

$$\phi^{(\text{eff})}(r) \;=\; -\frac{GM}{r} - \frac{GMh^2}{c^2r^3}$$

and their stability is determined by the sign of its second derivative. Here we re–examine the same problem with

$$h \quad \rightarrow \hbar\sqrt{\ell(\ell+1)}$$

and we determine whether quantization shifts the stability boundary.

Upon quantization, the effective potential becomes

$$\phi_\ell^{(\mathrm{eff})}(r) \;=\; -\frac{GM}{r} - \frac{GM\,\hbar^2\,\ell(\ell+1)}{c^2 r^3}$$

The condition for a circular orbit at radius $r = r_\mathrm{c}$ is

$$\left.\frac{d\phi_\ell^{(\mathrm{eff})}}{dr}\right|_{r_\mathrm{c}} = 0$$

which yields

$$\frac{GM}{r_\mathrm{c}^2} + \frac{3GM\,\hbar^2\,\ell(\ell+1)}{c^2 r_\mathrm{c}^4} \;=\; \frac{\hbar^2\,\ell(\ell+1)}{m_0 r_\mathrm{c}^3}$$

This is the quantum-corrected balance between the central force and the centrifugal term.

10.5.2 Second–derivative (stability) condition

Stability requires

$$\left.\frac{d^2\phi_\ell^{(\mathrm{eff})}}{dr^2}\right|_{r_\mathrm{c}} > 0$$

which gives

$$\frac{2GM}{r_\mathrm{c}^3} + \frac{12GM\,\hbar^2\,\ell(\ell+1)}{c^2 r_\mathrm{c}^5} \;>\; \frac{3\hbar^2\,\ell(\ell+1)}{m_0 r_\mathrm{c}^4}$$

Using the circular–orbit condition to eliminate r_c, one finds (after straightforward algebra)

$$r_{\mathrm{c}} > \frac{3GM}{c^2}\left[1 + O\left(\frac{\hbar^2}{m_0^2 c^2 r_{\mathrm{c}}^2}\right)\right]$$

The correction term is suppressed by the ratio $\hbar/(m_0 c r_{\mathrm{c}})$—the ratio of the Compton wavelength to the orbital radius—which for any macroscopic orbit is fantastically tiny.

10.5.3 Exact classical stability boundary

To leading order in $\hbar$, the stability condition therefore remains

$$r_{\mathrm{c}} > \frac{3GM}{c^2}$$

the same as in the classical Flat Space Gravity and numerically equal to the value that appears in GR for the innermost stable circular orbit (ISCO) of a non–rotating source.

10.5.4 Interpretation

Quantum discreteness of *h does* not destabilize circular FSG orbits. This is expected:

- The ratio $\hbar/(m_0 c r)$ is $\sim 10^{-40}$–10^{-50} for planetary and astrophysical orbits.
- The effective potential depends on h only via the $1/r^3$ term, making the quantum correction scale as $\hbar^2/r^4$.
- Stability is governed by derivatives of $\phi^{(\mathrm{eff})}$, which suppress the correction even further to $\sim \hbar^2/r^5$.

Thus, the quantum Flat Space Gravity predicts that *all classical stable orbits remain stable quantum mechanically*, and no new instability window appears.

10.5.5 Significance

This reinforces a recurring theme of this section:

- classical FSG predictions dominate entirely at macroscopic scales, - quantum corrections are beautifully structured but *suppressed by extreme powers of the orbital length scale*, - the structure of flat–space gauge propagation is preserved exactly.

The quantum stability analysis therefore closes the loop between the quantized field, the quantized orbit, and the classical limit, and confirms the internal coherence of the Flat Space Gravity framework from bound orbits to relativistic propagation.

10.6 Quantum–Corrected Energy Levels

10.6.1 Radial Schrödinger equation

A natural application of the quantized Flat Space Gravity framework is the determination of bound–state energy levels for a test particle of rest mass m_0 moving in the effective potential

$$\phi^{(\text{eff})}(r) \;=\; -\frac{GM}{r} - \frac{GMh^2}{c^2r^3}$$

with $h \to \hbar\sqrt{\ell(\ell+1)}$ after quantization. The first term is Coulomb–like and reproduces the Newtonian spectrum, while the second term is a genuinely relativistic FSG correction analogous to the spin–orbit and fine–structure terms familiar from hydrogen–like systems. Here we determine the leading quantum shift in the energy spectrum.

In the nonrelativistic limit ($v \ll c$), and for notational clarity, define the effective potential

$$V_\ell(r) \;=\; m_0\phi_\ell^{(\text{eff})}(r) \;=\; -\frac{GMm_0}{r} - \frac{GMm_0\hbar^2\ell(\ell+1)}{c^2r^3}$$

Then the radial Schrödinger equation is

$$-\frac{\hbar^2}{2m_0}\frac{d^2u}{dr^2}+\left[V_\ell(r)+\frac{\hbar^2\,\ell(\ell+1)}{2m_0r^2}\right]u(r) = E_{n\ell}\,u(r)$$

where $u(r) = rR(r)$ is the standard reduced radial wavefunction.

10.6.2 Zeroth–order spectrum (Newtonian)

Ignoring the $1/r^3$ term, the Coulombic potential produces the bound–state energies

$$E_n^{(0)} = -\frac{G^2M^2m_0^3}{2\hbar^2n^2}, \qquad n = 1,2,\ldots$$

identical in form to the hydrogen–like spectrum with gravitational coupling GMm_0.

10.6.3 First–order FSG correction

Treat the $1/r^3$ term as a perturbation:

$$H'_\ell = -\frac{GMm_0\hbar^2\ell(\ell+1)}{c^2r^3}$$

The first–order shift is

$$\Delta E_{n\ell} = \langle n\ell m|\,H'_\ell\,|n\ell m\rangle = -\frac{GMm_0\hbar^2\ell(\ell+1)}{c^2}\langle\frac{1}{r^3}\rangle_{n\ell}$$

For hydrogenic wavefunctions,

$$\left\langle\frac{1}{r^3}\right\rangle_{n\ell} = \frac{1}{n^3a_0^3}\frac{1}{\ell(\ell+1/2)(\ell+1)}$$

$$a_0 = \frac{\hbar^2}{GMm_0^2}$$

so that

$$\Delta E_{n\ell} = -\frac{GMm_0\hbar^2\ell(\ell+1)}{c^2}\frac{(GMm_0^2)^3}{\hbar^6}\frac{1}{n^3}\frac{1}{\ell(\ell+1/2)(\ell+1)}$$

After simplification the $\ell(\ell+1)$ factors cancel:

$$\Delta E_{n\ell} = -\frac{G^4M^4m_0^7}{c^2\hbar^4}\frac{1}{n^3\,(\ell+1/2)}$$

10.6.4 Magnitude of the shift

For any realistic gravitationally bound system, the dimensionless ratio

$$\epsilon = \frac{GMm_0}{\hbar c}$$

is absurdly small ($\epsilon \ll 10^{-40}$ for macroscopic masses), so the correction $\Delta E_{n\ell}$ is suppressed by ϵ^4:

$$\frac{\Delta E_{n\ell}}{E_n^{(0)}} \sim \epsilon^4 \ll 10^{-160}$$

Thus, the Flat Space Gravity quantum correction is entirely negligible in magnitude, as expected for macroscopic gravitational systems.

10.6.5 Interpretation

Although too small ever to be observed, this correction has genuine conceptual significance:

- It arises automatically from the *flat–space* FSG dynamics.
- The correction scales as $1/r^3$, confirming that the same term driving the classical perihelion shift also produces quantum fine–structure–like effects.
- The perturbation is structurally analogous to spin–orbit, Darwin, or relativistic corrections in hydrogen, but fully within flat spacetime and without metric curvature.

- The energy spectrum remains discrete and stable, with corrections that respect Flat Space Gravity gauge invariance.

10.6.6 Significance

Quantum bound–state structure in the Flat Space Gravity mirrors the structure found in conventional gauge theories: classical Coulombic levels receive minute but beautifully organized fine–structure corrections governed by the underlying $1/r^3$ interaction.

10.7 Quantum Gravitational Waves

10.7.1 Quantized transverse modes

In the previous section we established that the Flat Space Gravity field admits *exactly two transverse propagating degrees of freedom*, satisfying a wave equation formally identical to Maxwell's:

$$\Box \mathbf{g}_{\perp} = 0, \qquad \Box \mathbf{b}_{\perp} = 0$$

with no curvature, no weak–field approximations, and no perturbation of the metric. Here we complete the quantum picture by identifying the *quantum gravitational–wave sector* of the Flat Space Gravity, its polarization structure, its propagation, and the physical content of its quanta.

Once the gauge is fixed (e.g. Lorenz gauge), the Flat Space Gravity field reduces to two transverse dynamical components, exactly like QED. For each wave–vector **k** and polarization $\lambda = 1{,}2$:

$$\hat{G}_{\mu\nu}(x) =$$

$$\sum_{\lambda=1}^{2} \int \Box \frac{d^3k}{(2\pi)^3\, 2\omega_{\boldsymbol{k}}} \left[\epsilon_{\mu\nu}^{(\lambda)}(\boldsymbol{k})\, a_{\boldsymbol{k},\lambda}\, e^{-ik\cdot x} + \epsilon_{\mu\nu}^{(\lambda)}(\boldsymbol{k})^{*}\, a_{\boldsymbol{k},\lambda}^{\dagger}\, e^{+ik\cdot x}\right]$$

The polarization tensors $\epsilon_{\mu\nu}^{(\lambda)}$ satisfy

$$k^{\mu}\epsilon_{\mu\nu}^{(\lambda)} = 0, \qquad \epsilon_{\mu}^{(\lambda)\mu} = 0, \qquad \epsilon_{\mu\nu}^{(\lambda)} G^{\mu\nu} = 0$$

Thus, the quanta of the field are *massless spin–1 excitations*, the natural counterpart of photons, but representing fluctuations of the gravitational–acceleration field rather than of the electromagnetic field.

10.7.2 Propagation and energy transport

For each quantum mode:

$$\omega_{\mathbf{k}} = c\,|\mathbf{k}|, \qquad \hat{H}_{\text{field}} = \sum_{\lambda=1}^{2} \int d^3k\; \hbar\omega_{\mathbf{k}} \left(a_{\mathbf{k},\lambda}^{\dagger} a_{\mathbf{k},\lambda} + \frac{1}{2}\right)$$

Thus, gravitational radiation in the Flat Space Gravity:

- travels at the speed of light,
- is transverse,
- carries positive energy density $u_{\text{gw}} = \frac{c^2}{8\pi G}(\mathbf{b}^2 + \mathbf{g}^2/c^2)$
- obeys the Poynting–like FSG flux $\mathbf{S} = \frac{c^2}{4\pi G}\,\mathbf{g} \times \mathbf{b}$

These properties follow directly from the *linear flat–space wave equation*, without curvature or Ricci tensors.

10.7.3 Emission by accelerated masses

For a slowly varying source the radiated power has the same schematic form as Larmor radiation in electromagnetism,

$$P_{\text{rad}} = \frac{G}{3c^3}\langle \ddot{\mathbf{d}}^2 \rangle$$

where $\mathbf{d}$ is the “gravitational dipole moment'' of the mass distribution. Because isolated two–body systems have vanishing monopole and dipole radiation in the center–of–mass frame, the dominant contribution comes from the FSG quadrupole–like moment,

$$Q_{ij} = \int \rho(\mathbf{x}) \left(x_i x_j - \frac{1}{3}\, \delta_{ij}\, \mathbf{x}^2 \right) d^3x$$

leading to a power formula structurally parallel to the GR quadrupole formula, but derived entirely in flat spacetime:

$$P_{\mathrm{FSG}} = \frac{G}{5c^5} \langle \dddot{Q}_{ij}\, \dddot{Q}^{ij} \rangle$$

10.7.4 Quantum interpretation

Each radiated quantum has energy $\hbar\omega$ and carries one unit of gravitational "polarization" (helicity ± 1). Because the Flat Space Gravity field is Abelian, there are no graviton–self–interaction terms and the radiation remains linear even at high intensities, in contrast to GR where nonlinearities entwine the field with its own stress–energy tensor.

10.7.5 Consistency checks

The FSG gravitational–wave sector satisfies:

- linear superposition,
- exact gauge invariance in flat spacetime,
- two transverse polarizations,
- positive–energy quanta,
- identical vacuum structure to QED,
- conserved stress–energy tensor,
- well–defined renormalization at the quantum level.

10.7.6 Significance

Quantum gravitational waves in the Flat Space Gravity provide a dramatic synthesis:

- *All classical wave properties* (speed c, transverse polarization, quadrupole emission, energy flux) arise *without* curved geometry.

- *All quantum features* (creation/annihilation operators, discrete quanta, vacuum polarization, radiation spectra) follow exactly the pattern of a massless Abelian gauge field in flat spacetime.

- The resulting theory is both *computationally simple* and *conceptually transparent*, avoiding the nonlinear graviton self-couplings inherent in GR.

10.8 The Flat–Space Consistency of Quantum FSG

10.8.1 Dimensional analysis and power counting

A final structural ingredient of the quantized Flat Space Gravity concerns the *renormalization programme*. Since the Flat Space Gravity field is an Abelian massless gauge field in flat Minkowski space, its quantum corrections follow the same schematic pattern as QED: vacuum polarization, vertex corrections, and particle self–energies. However, the dimensionality of the FSG coupling and the absence of metric nonlinearities offer profound simplifications.

In Flat Space Gravity the effective quantum coupling κ^2 has dimensions of (length)^2 in natural units. Thus, the coupling constant that appears in Feynman rules is effectively Gm_0^2 (for a particle of mass m_0). By standard power–counting arguments:

- The superficial degree of divergence of any FSG diagram is *no worse* than in QED.
- Vacuum polarization diagrams are logarithmically divergent.

- Vertex corrections and self–energies are at most logarithmically divergent.

The key point is that the Flat Space Gravity field is Abelian and *linear*: there are no 3– or 4–graviton vertices, no field self–couplings, and no curvature–driven nonlinearities. Hence the renormalization structure is exceptionally clean.

10.8.2 Renormalized propagator

After renormalization, the propagator retains exactly the same transverse projector:

$$D_{\mu\nu}^{(\mathrm{ren})}(k) \;=\; \frac{-\,\eta_{\mu\nu}^{\perp}(k)}{k^2\,[1-\Pi^{(\mathrm{ren})}(k^2)]}$$

with the renormalized vacuum–polarization function $\Pi^{(\mathrm{ren})}(k^2)$ satisfying

$$k^{\mu}\Pi_{\mu\nu}^{(\mathrm{ren})}(k) \;=\; 0$$

This ensures the strict maintenance of gauge invariance and the absence of unphysical longitudinal modes.

10.8.3 Running of the effective coupling

The effective FSG coupling becomes momentum–dependent:

$$G_{\mathrm{eff}}(k^2) \;=\; \frac{G}{1-\Pi^{(\mathrm{ren})}(k^2)}$$

Because the FSG current is always conserved,

$$k_{\mu}J^{\mu} = 0$$

the Ward identity holds identically, and therefore:

- $\Pi^{(\mathrm{ren})}(0) = 0$ by the renormalization condition,

- $G_{\text{eff}}(k^2) \to G$ in the infrared,
- running effects enter only at extremely small distances and are suppressed by powers of $k^2c^2/m_0^2c^4$.

For any macroscopic gravitational phenomenon ($k \sim 1/b$ with b an astronomical impact parameter),

$$G_{\text{eff}}(k^2) \;=\; G\,[\,1 + \mathcal{O}(10^{-60})\,]$$

10.8.4 Absence of strong–coupling or Landau–pole behavior

Because the renormalization group equation has the schematic structure

$$k^2 \frac{dG_{\text{eff}}}{dk^2} \;\propto\; G_{\text{eff}}^2\, k^2$$

the effective coupling does *not* diverge at any finite scale. The Flat Space Gravity therefore avoids the ultraviolet Landau pole that plagues QED, simply because G carries positive mass dimension. The theory is smoothly behaved in the deep ultraviolet and automatically suppresses quantum corrections at macroscopic scales.

10.8.5 Renormalized vertex and current conservation

The renormalized vertex satisfies

$$\Gamma_\mu^{(\text{ren})}(p',p) \;=\; U_\mu + \delta\Gamma_\mu(p',p)$$
$$k^\mu \Gamma_\mu^{(\text{ren})}(p',p) = 0$$

ensuring that:

- the renormalized charge (or coupling) remains gauge–invariant,
- longitudinal contributions remain absent to all orders,
- physical amplitudes involve only the two transverse helicity states.

10.8.6 Finite physical predictions

Every observable calculated in the quantized Flat Space Gravity —scattering, light deflection, radiation emission, bound–state shifts— is finite after renormalization, while classical predictions (perihelion, light deflection, redshift, Shapiro delay) remain intact. No curvature, no Ricci tensor, and no weak–field approximations are needed: everything is formulated directly in flat spacetime.

10.8.7 Significance

The renormalization analysis reinforces a central lesson in this section:

- The quantized Flat Space Gravity field behaves like a *perfectly consistent, massless, Abelian gauge theory* in flat space.

- Quantum corrections are structurally clean, UV–benign, and analytically tractable.

- Classical predictions dominate at all observable scales, with quantum modifications entering only as minuscule renormalizations.

- The Flat Space Gravity framework achieves a complete and self–contained quantum formulation without ever invoking geometric curvature or nonlinear field self–interactions.

10.9 Semiclassical–Quantum Bridge

10.9.1 Operator form of the orbital potential

Throughout the previous sections we saw that the *effective* potential

$$\phi^{(\mathrm{eff})}(r) \;=\; -\frac{GM}{r} - \frac{GMh^2}{c^2 r^3}$$

plays a decisive dynamical role: it governs perihelion precession, light deflection, and the existence of stable circular orbits. The orbital potential reappeared in the Hamiltonian through the canonical substitution

$$h \;\rightarrow\; \hbar\sqrt{\ell(\ell+1)}$$

revealing that the entire structure of celestial mechanics can be viewed as a smooth semiclassical–quantum hierarchy. Here we make this bridge explicit and formulate the precise quantum operator that corresponds to the classical effective potential.

Formally promoting h^2 to the operator

$$\hat{h}^2 \;=\; -\hbar^2\left[\frac{1}{\sin\theta}\frac{\partial}{\partial\theta}(\sin\theta\,\frac{\partial}{\partial\theta}) + \frac{1}{\sin^2\theta}\frac{\partial^2}{\partial\varphi^2}\right]$$

the effective potential becomes the Hermitian operator

$$\widehat{\phi^{(\mathrm{eff})}} \;=\; -\frac{GM}{r} - \frac{GM}{c^2r^3}\,\hat{h}^2$$

Acting on any spherical harmonic $Y_{\ell m}(\theta,\varphi)$,

$$\hat{h}^2\,Y_{\ell m} \;=\; \hbar^2\,\ell(\ell+1)\,Y_{\ell m}$$

and therefore

$$\widehat{\phi^{(\mathrm{eff})}}\,Y_{\ell m} \;=\; \left[-\frac{GM}{r} - \frac{GM\hbar^2\ell(\ell+1)}{c^2r^3}\right]Y_{\ell m}$$

10.9.2 Semiclassical limit

Using WKB wavefunctions with action variable $J_\varphi = h$ and quantization rule $J_\varphi = \hbar(\ell + 1/2)$, we recover the classical potential:

$$\phi^{(\mathrm{eff})}(r) \;=\; -\frac{GM}{r} - \frac{GMh^2}{c^2r^3} + O(\hbar^2)$$

which confirms that the FSG orbital term has an unambiguous and smooth quantum limit.

10.9.3 Quantum corrections

Using the operator form we can read off the leading $\hbar$–dependent modifications:

- Corrections $\propto \hbar^2/r^3$ in the potential,
- Corrections $\propto \hbar^2/r^4$ in the precession frequency,
- Corrections $\propto \hbar^2/r^5$ in the orbital stability condition,
- Corrections $\propto \hbar^2/r^3$ in the bound–state spectrum,
- Corrections $\propto G_{\text{eff}}(k^2)/G$ through radiative effects.

All of these scale with enormous inverse powers of the macroscopic radius r and are therefore unmeasurably small, but their *structure* is important: Flat Space Gravity reproduces the correct classical dynamics and gives the expected minimal quantum deformations.

10.9.4 Bridge to classical trajectories

The expectation value of the effective potential in any WKB state yields directly the classical equation for the orbit:

$$\left\langle \widehat{\phi^{(\text{eff})}} \right\rangle = -\frac{GM}{r} - \frac{GMh^2}{c^2 r^3} + \mathcal{O}(\hbar^2)$$

and hence

$$\frac{d^2u}{d\varphi^2} + u = \frac{GM}{h^2} + \frac{3GM}{c^2} u^2 + \mathcal{O}(\hbar^2)$$

which is the precise FSG analogue of the Binet equation governing the perihelion advance and the bending of trajectories.

10.9.5 Significance

This final subsection crystallizes the structure that has been emerging throughout previous sections:

- The effective orbital potential has a clean operator meaning.
- Quantum mechanics modifies it in exactly the expected way.
- Its expectation value reproduces the classical FSG dynamics.
- Its fluctuations generate minuscule but well–defined quantum shifts.
- The whole construction is completed in *flat spacetime*, with no curvature and no geometric perturbation theory.

Thus, the Flat Space Gravity framework achieves a rare synthesis: *a classical–quantum dictionary for orbital dynamics* that is fully self–consistent, gauge–invariant, and constructed entirely in the language of flat–space field theory.

10.10 Concluding Remarks

In this section the fully quantum structure of the Flat Space Gravity has been constructed and analyzed in detail. Canonical quantisation of bound orbits yields discrete energy spectra and well–defined quantum corrections to classical observables such as the perihelion shift and light deflection. Orbital stability emerges naturally from the operator formulation of the effective potential, while radiative processes are described in terms of quantized normal modes of the gravitational gauge field. Renormalization and running–coupling effects were shown to remain ultraviolet benign, preserving exact gauge consistency and ensuring finite physical predictions at all measurable scales. The quantum formulation therefore remains analytically tractable and free from strong–coupling pathologies, owing to the linear, Abelian nature of the Flat Space Gravity field and to the formulation of the theory on flat spacetime.

Together with the preceding semiclassical developments, the results presented here establish a complete classical–to–quantum continuum for gravitational dynamics within the Flat Space Gravity framework. The theory achieves a coherent synthesis of orbital mechanics, radiation, and quantum field behavior without invoking spacetime curvature or non–linear geometric structures, thereby providing a self–consistent quantum alternative to conventional geometric approaches.

11. Cosmology in Flat Spacetime Gravity

11.1 Introductory Remark

In the previous sections we developed Flat Spacetime Gravity as a fully relativistic field theory formulated on a Minkowski background. The antisymmetric tensor $G_{\mu\nu}$, together with its sourceful and rotational projections, provided a complete description of local gravitational dynamics, test–particle motion, radiative phenomena and canonical structures. Astronomical tests such as perihelion precession, light deflection, redshift and Shapiro delay were all recovered within flat spacetime through the effective potential $\phi^{(\text{eff})}$ and the associated dynamical law $a = G^{(\text{eff})}U$.

In this final section we turn to the largest possible scale: namely the cosmological domain. The central question is whether a theory that is strictly built on flat spacetime geometry can still account for the observed large–scale behavior of the Universe, including expansion, horizon structure, cosmological redshift and the inferred mass distribution. Rather than introducing a curved metric, Flat Spacetime Gravity treats cosmology as a problem of organizing the sourceful and rotational flat–spacetime gravitational fields generated by a homogeneous and possibly time–dependent matter distribution.

The strategy is the following. We first formulate a cosmological ansatz for the Flat Spacetime Gravity fields compatible with large–scale homogeneity and isotropy at the level of the sourceful field and with possible global rotational components encoded in the rotational field. We then derive the corresponding evolution equations for the effective potential and for the average energy density, and show how a cosmological redshift arises from the same γ–structure that already governed local gravitational redshift in the previous sections. Finally, we discuss to what extent flat–spacetime dynamics can reproduce the observed phenomenology usually attributed to curved cosmological metrics, and we outline the implications of Flat Spacetime Gravity for large–scale structure and the interpretation of dark matter and dark energy.

11.2 Cosmological Ansatz for Flat Spacetime Gravity

A cosmological description within Flat Spacetime Gravity must be consistent with the two empirically established large–scale properties of the Universe:

- *homogeneity* of the average matter distribution,
- *isotropy* around any comoving observer.

Since Flat Spacetime Gravity is formulated on a strictly Minkowskian background $\eta_{\mu\nu} = \mathrm{diag}(-1, +1, +1, +1)$, homogeneity and isotropy cannot be implemented by curvature of spacetime, but only through the structure of the sourceful and rotational gravitational fields generated by a spatially uniform matter density.

We therefore introduce the following cosmological ansatz:

$$\rho_0 = \rho_0(t), \qquad U^\mu = (c,\, 0,\, 0,\, 0)$$

where $\rho_0(t)$ is the rest–mass density averaged over sufficiently large comoving volumes, and U^μ is the four–velocity of a comoving observer. A homogeneous distribution cannot generate a net rotational field on average, so the leading–order ansatz is

$$\mathbf{b}(t, \mathbf{x}) = \mathbf{0}$$

while the sourceful field $\mathbf{g}$ is necessarily radial and may be time–dependent:

$$\mathbf{g}(t, \mathbf{x}) = g_r(t, r)\, \mathbf{e}_r$$

In terms of the FSG potential this corresponds to a purely temporal potential,

$$\Phi_\mu = \left(-\frac{\phi(t, r)}{c},\ 0{,}0{,}0\right)$$

whose spatial gradient produces the radial sourceful field

$$\mathbf{g} = -\nabla\phi(t,r)$$

The structure above is the flat–spacetime analogue of a Friedmann–type cosmological configuration: homogeneity imposes that all dynamical information relevant to cosmic evolution enters only through the time dependence of $\rho_0(t)$ and of the radial potential $\phi(t,r)$, while isotropy forbids any privileged spatial direction.

This ansatz allows the full FSG equations to be reduced to a small set of ordinary differential relations governing $\phi(t,r)$, the cosmological redshift factor $\gamma(t)$, and the evolution of the average energy density. These relations will be developed in the following subsections.

11.3 Field Equations for a Homogeneous–Isotropic Distribution

With the cosmological ansatz of this section, the full Flat Spacetime Gravity field tensor

$$G_{\mu\nu} = \partial_\mu\Phi_\nu - \partial_\nu\Phi_\mu$$

reduces to a single independent component: the radial sourceful field

$$\mathbf{g}(t,r) = -\nabla\phi(t,r), \qquad \mathbf{b}(t,r) = \mathbf{0}$$

Since the four–current of a comoving homogeneous medium is

$$J^\mu = \rho_0(t)U^\mu = (\rho_0(t)c, 0,0,0)$$

the inhomogeneous Maxwell–type Flat Spacetime Gravity equation,

$$\partial_\mu G^{\mu\nu} = -\frac{4\pi G}{c^2} J^\nu$$

yields immediately, for the purely temporal component $\nu = 0$,

$$\partial_i G^{i0} = -\frac{4\pi G}{c^2}\rho_0 c$$

and since $G^{i0} = g^i/c$ we obtain the flat–spacetime Poisson equation:

$$\nabla \cdot \mathbf{g}(t,r) = -4\pi G \rho_0(t)$$

For a spherically symmetric potential this becomes

$$\frac{1}{r^2}\frac{\partial}{\partial r}(r^2 g_r(t,r)) = -4\pi G \rho_0(t)$$

Integrating with respect to r

$$g_r(t,r) = -\frac{4\pi G}{3}\rho_0(t)\,r + \frac{F(t)}{r^2}$$

We set F(t) = 0 for normalization reasons at $r = 0$, so

$$g_r(t,r) = -\frac{4\pi G}{3}\rho_0(t)\,r$$

which in turn integrates to the cosmological potential

$$\phi(t,r) = +\frac{2\pi G}{3}\rho_0(t)\,r^2 + C(t)$$

(up to an arbitrary additive function $C(t)$ that can be absorbed into the definition of the redshift factor $\gamma(t)$).

Thus, the Flat Spacetime Gravity field equations imply that a spatially uniform rest–mass density produces a *quadratic* gravitational potential:

$$\phi \propto \rho_0\, r^2$$

exactly as expected for a homogeneous mass distribution in flat spacetime.

In the homogeneous–isotropic Flat Spacetime Gravity cosmology the scalar potential obtained from the Poisson–type equation

$$\nabla \cdot \mathbf{g}(t,\mathbf{r}) = -4\pi G\,\rho_0(t), \qquad \mathbf{g} = -\nabla\phi$$

Reads

$$\phi(t,r) = \frac{2\pi G}{3}\rho_0(t)\,r^2$$

which is the familiar interior solution for a homogeneous sphere. This $\phi(t,r)$ should be viewed as the *geometric* (Poisson) potential that reproduces the correct field

$$\mathbf{g}(t,r) = -\nabla\phi(t,r) = -\frac{4\pi G}{3}\rho_0(t)\, r\,\hat{\mathbf{r}}$$

$$\nabla\cdot\mathbf{g} = -4\pi G\,\rho_0(t) < 0$$

However, the Lorentz factor $\gamma(t,r)$ in Flat Spacetime Gravity couples not to this Poisson potential itself, but to the *gravitational potential energy per unit mass*. For a homogeneous sphere one has

$$\mathcal{U}(t,r) = \frac{1}{2}\,\mathbf{g}(t,r)\cdot\mathbf{r} = -\frac{2\pi G}{3}\rho_0(t)\, r^2 = -\phi(t,r)$$

It is this quantity $\mathcal{U}(t,r)$ that enters the time–rate factor. For clarity we introduce a cosmological potential

$$\phi_{\text{cosm}}(t,r) \equiv \mathcal{U}(t,r) = -\phi_{\text{Poisson}} = -\phi(t,r)$$

thus,

$$\phi_{\text{cosm}}(t,r) = -\frac{2\pi G}{3}\rho_0(t)\, r^2$$

so that the FSG relation between γ and the potential keeps its universal form

$$\gamma(t,r) = 1 - \frac{\phi_{\text{cosm}}(t,r)}{c^2} = 1 + \frac{2\pi G}{3c^2}\rho_0(t)\, r^2$$

We emphasize that the sourceful field is always obtained from $\mathbf{g}(t,r) = -\nabla\phi(t,r)$, whereas the time-rate factor couples to the cosmological potential $\phi_{\text{cosm}}(t,r)$ rather than directly to the Poisson potential $\phi(t,r)$.

The temporal evolution of $\rho_0(t)$ enters only parametrically at this stage; dynamical consequences such as cosmological redshift and Hubble–type expansion arises when this potential is inserted into the FSG dynamical law

$$a^{\mu} = \mathrm{G}^{(\text{eff})\mu}_{\nu}\, U^{\nu}$$

The next subsection develops precisely this connection.

11.4 Cosmological Redshift in Flat Spacetime Gravity

In the previous sections we showed that, for a test particle moving in the effective potential $\phi^{(\text{eff})}$, the dynamical law

$$a^{\mu} = G^{(\text{eff})\mu}{}_{\nu}\, U^{\nu}$$

implies an energy relation of the form

$$\gamma = 1 - \frac{\phi^{(\text{eff})}}{c^2} + \frac{E}{c^2}$$

where E is a constant of motion associated with time translation invariance and γ is the usual relativistic factor. In the cosmological case considered here we have no orbital contribution, so

$$\phi^{(\text{eff})}(t,r) = \phi_{\text{cosm}}(t,r)$$

with $\phi_{\text{cosm}}(t,r)$ given by the homogeneous solution of this section:

$$\phi_{\text{cosm}}(t,r) = -\frac{2\pi G}{3}\rho_0(t)\,r^2 + C(t)$$

For massive test particles comoving with the homogeneous medium ($U^{\mu} = (c,0,0,0)$) the effective potential determines the way in which their local energy and clock rate differ from those of a freely falling inertial observer at infinity. For lightlike signals, the same γ structure governs how photon frequencies are measured by comoving observers at different events in spacetime.

Consider a photon emitted at an event $(t_{\text{em}}, r_{\text{em}})$ and received by a comoving observer at $(t_{\text{obs}}, r_{\text{obs}})$. The frequency measured by such an observer is proportional to the local time rate associated with $\gamma(t,r)$. To leading order in ϕ_{cosm}/c^2 we may write

$$\gamma(t,r) \simeq 1 - \frac{\phi_{\rm cosm}(t,r)}{c^2}$$

for a photon with vanishing rest mass and with the integration constant E set to zero, so all frequency shift is encoded in the potential term. The ratio of observed to emitted frequencies is then

$$\frac{\nu_{\rm obs}}{\nu_{\rm em}} \simeq \frac{\gamma(t_{\rm obs}, r_{\rm obs})}{\gamma(t_{\rm em}, r_{\rm em})} \simeq \frac{1 - \phi_{\rm cosm}(t_{\rm obs}, r_{\rm obs})/c^2}{1 - \phi_{\rm cosm}(t_{\rm em}, r_{\rm em})/c^2}$$

For weak fields we expand to first order in ϕ/c^2:

$$\frac{\nu_{\rm obs}}{\nu_{\rm em}} \simeq 1 + \frac{\phi_{\rm cosm}(t_{\rm em}, r_{\rm em}) - \phi_{\rm cosm}(t_{\rm obs}, r_{\rm obs})}{c^2}$$

so that the cosmological redshift

$$z \equiv \frac{\nu_{\rm em} - \nu_{\rm obs}}{\nu_{\rm obs}}$$

is approximately

$$z \simeq \frac{\phi_{\rm cosm}(t_{\rm obs}, r_{\rm obs}) - \phi_{\rm cosm}(t_{\rm em}, r_{\rm em})}{c^2}$$

In FSG cosmology the potential is

$$\phi_{\rm cosm}(t,r) = -\frac{2\pi G}{3} \rho_0(t)\, r^2 + C(t)$$

so, for a comoving galaxy at fixed radius r_* we have

$$\phi_{\rm cosm,em} = -\frac{2\pi G}{3} \rho_0(t_{\rm em})\, r_*^2 + C(t_{\rm em})$$
$$\phi_{\rm cosm,obs} = -\frac{2\pi G}{3} \rho_0(t_{\rm obs})\, r_*^2 + C(t_{\rm obs})$$

A suitable choice of the time–dependent offset $C(t)$ allows one to absorb the purely temporal contribution into the global definition of the time coordinate or of the energy constant, leaving the spatially dependent part

$$\Delta\phi_{\text{cosm}} \simeq \frac{2\pi G}{3}\left[\rho_0(t_{\text{em}}) - \rho_0(t_{\text{obs}})\right] r_*^2$$

Thus, within Flat Spacetime Gravity, cosmological redshift arises not from the stretching of a curved spacetime metric but from the evolution of the flat–spacetime potential $\phi(t,r)$ and the associated γ factor along the photon worldline. The same kinematical structure that produced the local gravitational redshift in the previous sections now generates a global, scale–dependent redshift in a homogeneous Universe, entirely within flat spacetime.

11.5 Hubble–Type Expansion in Flat Spacetime

11.5.1 Effective recession velocity from $\gamma(t,r)$

We now analyze how a Hubble–type expansion arises in Flat Spacetime Gravity purely through the time dependence of the local time–rate factor

$$\gamma(t,r) = 1 - \frac{\phi_{\text{cosm}}(t,r)}{c^2}$$

$$\phi_{\text{cosm}}(t,r) = -\frac{2\pi G}{3}\rho_0(t)\, r^2 + C(t)$$

with the corresponding sourceful field

$$g_r(t,r) = -\frac{\partial\phi}{\partial r} = -\frac{4\pi G}{3}\rho_0(t)\, r$$

In Flat Spacetime Gravity the geometry is flat, so "expansion" cannot be encoded in the metric. Instead, physical separations between comoving worldlines evolve because $\gamma(t,r)$ evolves in time. We show this explicitly.

Let two galaxies be comoving and separated by a fixed FSG coordinate distance r. Define an effective scale factor

$$a_{\text{FSG}}(t,r) \equiv \frac{\gamma(t,r)}{\gamma(t_0,r)}$$

Then the physical separation is

$$\ell(t,r) = a_{\mathrm{FSG}}(t,r)\,r$$

The effective recession velocity follows by differentiation:

$$v_{\mathrm{eff}}(t,r) \equiv \frac{d\ell}{dt}$$
$$= r\,\frac{d}{dt}\left(\frac{\gamma(t,r)}{\gamma(t_0,r)}\right) = r\,\frac{\dot{\gamma}(t,r)}{\gamma(t_0,r)}$$

Since $\gamma(t,r) \simeq \gamma(t_0,r)$ for small ϕ/c^2,

$$v_{\mathrm{eff}}(t,r) \simeq r\,\frac{\dot{\gamma}(t,r)}{\gamma(t,r)} = r\,\frac{d}{dt}(\ln\gamma(t,r))$$

Hence the *effective Hubble law* in flat spacetime is

$$v_{\mathrm{eff}}(t,r) = r\,\frac{d}{dt}\big(\ln\gamma(t,r)\big)$$

The corresponding effective acceleration is

$$a_{\mathrm{eff}}(t,r) \equiv \frac{dv_{\mathrm{eff}}}{dt} = r\,\frac{d^2}{dt^2}\big(\ln\gamma(t,r)\big)$$

which has the correct physical dimensions.

11.5.2 Explicit derivation of a_{eff} and v_{eff}

We compute $\dot{\gamma}$:

$$\gamma(t,r) = 1 - \frac{\phi_{\mathrm{cosm}}(t,r)}{c^2}$$
$$\dot{\gamma}(t,r) = -\frac{1}{c^2}\frac{\partial\phi_{\mathrm{cosm}}}{\partial t}$$

Using

$$\phi_{\mathrm{cosm}}(t,r) = -\frac{2\pi G}{3}\rho_0(t)\,r^2 + C(t)$$

we obtain

$$\frac{\partial\phi_{\text{cosm}}}{\partial t} = -\frac{2\pi G}{3}r^2\dot{\rho}_0(t) + \dot{C}(t)$$

As usual, the time–only term $\dot{C}(t)$ may be absorbed into a redefinition of the local clock or the constant of motion; only the r^2 part contributes to cosmological behaviour. Thus,

$$\dot{\gamma}(t,r) \simeq +\frac{2\pi G}{3c^2}r^2\dot{\rho}_0(t)$$

To leading order,

$$\frac{d}{dt}(\ln\gamma) = \frac{\dot{\gamma}}{\gamma} \simeq +\frac{2\pi G}{3c^2}r^2\dot{\rho}_0(t)$$

Therefore,

$$v_{\text{eff}}(t,r) \simeq +\frac{2\pi G}{3c^2}r^3\dot{\rho}_0(t)$$

$$a_{\text{eff}}(t,r) = \frac{dv_{\text{eff}}}{dt} \simeq +\frac{2\pi G}{3c^2}r^3\ddot{\rho}_0(t)$$

We emphasize:

$$v_{\text{eff}} \propto r^3\,\dot{\rho}_0(t), \qquad a_{\text{eff}} \propto r^3\,\ddot{\rho}_0(t)$$

so, velocity and acceleration correctly differ by one time derivative. The similarity of their functional form is merely because both are derivatives of the same basic object $\ln\gamma$.

11.5.3 Why the effective Hubble parameter can be treated as constant

Define the effective Hubble factor:

$$\text{H}_{\text{eff}}(t,r) \equiv \frac{v_{\text{eff}}(t,r)}{r}$$

$$= \frac{d}{dt}\left(\ln\gamma(t,r)\right) \simeq +\frac{2\pi G}{3c^2}r^2\dot{\rho}_0(t)$$

Although H_{eff} depends on both t and r, it can be treated as approximately constant when:

- the cosmic density varies slowly, i.e. $\dot{\rho}_0(t) \approx$ const over observational time scales;
- the observational sample spans distances small compared with the Flat Spacetime Gravity cosmological scale, so that the variation of r^2 across the sample is mild.

Under these conditions,

$$\mathrm{H}_{\text{eff}}(t, r) \;\simeq\; H_0$$

where H_0 corresponds to a representative distance r_* and the present cosmic epoch t_0. Consequently,

$$v_{\text{eff}}(t_0, r) \;\simeq\; H_0 r$$

recovering a Hubble–type law entirely within flat spacetime.

Thus, in FSG cosmology:

- expansion arises not from curvature but from time evolution of the local time–rate factor $\gamma(t, r)$;
- the recession velocity follows the exact identity

$$v_{\text{eff}}(t, r) \;=\; r\,\frac{d}{dt}(\ln\gamma(t, r))$$

- a linear Hubble law emerges naturally in the observational regime, without any curved spacetime.

11.5.4 Summary

The result of this subsection is remarkable in its simplicity and its conceptual clarity: a Hubble–type expansion law emerges in a strictly flat–spacetime setting, without any appeal to curvature, scale factors, or geometric expansion of space itself. The entire cosmological behavior originates from the slow temporal evolution of the local time–rate factor

$\gamma(t,r)$, which in Flat Spacetime Gravity encodes the influence of the evolving matter density on physical rulers, clocks, and recession velocities. In this framework, the familiar Hubble law becomes

$$v_{\text{eff}}(t,r) = r\frac{d}{dt}(\ln\gamma(t,r))$$

a purely kinematic consequence of flat geometry interacting with time–dependent gravitational potentials. The formal similarity in the functional form of $v_{\text{eff}}(t,r)$ and $a_{\text{eff}}(t,r)$ reflects the fact that both are time derivates of $\ln\gamma(t,r)$; their physical dimensions differ, as they involve first and second derivates $\rho_0(t)$ of respectively.

This demonstrates that flat spacetime is fully capable of supporting observationally relevant cosmological phenomenology, once the role of $\gamma(t,r)$ is properly understood and incorporated.

11.6 Matter–Density Evolution and the Effective Hubble Law

11.6.1 Continuity equation with r–dependent $\gamma(t,r)$

In Flat Spacetime Gravity the cosmic recession does not follow from a stretching of spatial geometry but from the time–evolution of the local gravitational time–rate factor

$$\gamma(t,r) = 1 - \frac{\phi_{\text{cosm}}(t,r)}{c^2}$$

$$\phi_{\text{cosm}}(t,r) = -\frac{2\pi G}{3}\rho_0(t)\,r^2$$

Because the potential scales as r^2, the recession velocities derived in previous sections are likewise r–dependent:

$$v_{\text{eff}}(t,r) = r\frac{d}{dt}[\ln\gamma(t,r)]$$

To determine the global behavior, the density evolution $\rho_0(t)$ must be solved from mass conservation in spherical symmetry.

In flat spacetime the continuity equation is

$$\frac{\partial \rho_0}{\partial t} + \nabla \cdot (\rho_0 \, \mathbf{v}_{\text{eff}}) = 0$$

with

$$\mathbf{v}_{\text{eff}}(t,r) = v_{\text{eff}}(t,r)\, \mathbf{e}_r = r \frac{\partial}{\partial t}[\ln\gamma(t,r)]\mathbf{e}_r$$

Using spherical symmetry,

$$\nabla \cdot (\rho_0 \mathbf{v}_{\text{eff}}) = \frac{1}{r^2}\frac{\partial}{\partial r}[r^2 \rho_0(t) v_{\text{eff}}(t,r)]$$

Since ρ_0 depends only on t, while v_{eff} depends on r through $\gamma(t,r)$, we obtain

$$\nabla \cdot (\rho_0 \mathbf{v}_{\text{eff}}) = \rho_0(t)\left[3\frac{\partial}{\partial t}(\ln\gamma(t,r)) + r\frac{\partial}{\partial r}\left(\frac{\partial}{\partial t}\ln\gamma(t,r)\right)\right]$$

From

$$\gamma(t,r) = 1 + \frac{2\pi G}{3c^2}\rho_0(t)\, r^2$$

we compute

$$\frac{\partial}{\partial r}\left(\frac{\partial}{\partial t}\ln\gamma\right) = +\frac{4\pi G}{3c^2} r \frac{d\rho_0}{dt}\frac{1}{\gamma(t,r)}$$

Therefore

$$\nabla \cdot (\rho_0 \mathbf{v}_{\text{eff}}) = 3\rho_0 \frac{\dot{\gamma}}{\gamma} + \rho_0 \frac{4\pi G}{3c^2} r^2 \frac{\dot{\rho}_0}{\gamma}$$

The continuity equation becomes

$$\dot{\rho}_0 + 3\rho_0 \frac{\dot{\gamma}}{\gamma} + \rho_0 \frac{4\pi G}{3c^2} r^2 \frac{\dot{\rho}_0}{\gamma} = 0$$

Solving for $\dot{\rho}_0$ gives

$$\dot{\rho_0} = -3\rho_0 \frac{\dot{\gamma}(t,r)}{\gamma(t,r)}\left[1 + \frac{4\pi G}{3c^2}\frac{\rho_0 r^2}{\gamma(t,r)}\right]^{-1}$$

To obtain the global (cosmological) limit, we evaluate this at the horizon scale $r = R$, for which $\gamma(t, R)$ remains positive as long as $\rho_0(t)$ is below the Flat Spacetime Gravity collapse threshold. Thus, the density evolution law is

$$\dot{\rho_0}(t) = -3\rho_0(t)\, H_{\text{eff}}(t, R)\left[1 + \frac{4\pi G}{3c^2}\frac{\rho_0 R^2}{\gamma(t, R)}\right]^{-1}$$

where

$$H_{\text{eff}}(t,r) = \frac{1}{\gamma(t,r)}\frac{\partial \gamma(t,r)}{\partial t}$$

11.6.2 Cosmological interpretation

In the weak–field regime ($\gamma \simeq 1$ and $2\pi G\rho_0 R^2/3c^2 \ll 1$), the square bracket becomes 1 and

$$\dot{\rho_0}(t) \simeq -3\rho_0(t)\, H_{\text{eff}}(t, R)$$

so that

$$\rho_0(t) \propto \exp\left(-3\int^t H_{\text{eff}}(t')\, dt'\right)$$

Hence the FSG cosmology recovers the familiar dilution law $\rho \propto a^{-3}$ but without any metric scale factor $a(t)$. Instead, the whole effect arises from the time–rate factor $\gamma(t, r)$ and its r–dependence.

11.6.3 Conceptual closure

In Flat Spacetime Gravity the density decreases not because space expands but because the local clock–rate $\gamma(t, r)$ accelerates as $\rho_0(t)$ falls. This mechanism induces recession velocities

$$v_{\text{eff}}(t,r) = H_{\text{eff}}(t,r)\,r$$

and a Hubble–like evolution law

$$H_{\text{eff}} = \frac{\dot{\gamma}}{\gamma}$$

in a purely Minkowskian background.

11.7 Effective Deceleration Parameter

11.7.1 Kinematic definition

In standard cosmology the deceleration parameter q encodes how the Hubble function changes in time. In Flat Spacetime Gravity, where recession arises from the time variation of the local gravitational time–rate factor $\gamma(t,r)$ rather than from metric expansion, we can define an analogous quantity directly in terms of the effective Hubble function

$$H_{\text{eff}}(t,r) \equiv \frac{1}{\gamma(t,r)}\frac{\partial\gamma(t,r)}{\partial t}$$

We introduce the effective deceleration parameter in the usual kinematic form

$$q_{\text{eff}}(t,r) \equiv -\frac{1}{H_{\text{eff}}^2(t,r)}\frac{\partial H_{\text{eff}}(t,r)}{\partial t}$$

A positive q_{eff} corresponds to decelerating recession, a negative q_{eff} to accelerating recession, and $q_{\text{eff}} = 0$ to a strictly exponential expansion rate.

With

$$H_{\text{eff}} = \frac{\dot{\gamma}}{\gamma}$$

we have

$$\dot{H}_{\text{eff}} = \frac{\ddot{\gamma}}{\gamma} - \frac{\dot{\gamma}^2}{\gamma^2}$$

so that

$$q_{\text{eff}} = -\frac{1}{H_{\text{eff}}^2}\left(\frac{\ddot{\gamma}}{\gamma} - \frac{\dot{\gamma}^2}{\gamma^2}\right) = 1 - \frac{\gamma\,\ddot{\gamma}}{\dot{\gamma}^2}$$

This compact expression is fully kinematic and does not refer to any geometry beyond the flat Minkowski background.

11.7.2 Expression in terms of the matter density

In the homogeneous–isotropic FSG cosmology of this section, the potential and time–rate factor are

$$\phi_{\text{cosm}}(t,r) = -\frac{2\pi G}{3}\,\rho_0(t)\,r^2, \qquad \gamma(t,r) = 1 - \frac{\phi_{\text{cosm}}(t,r)}{c^2}$$
$$= 1 + A\,\rho_0(t)\,r^2$$

where

$$A \equiv \frac{2\pi G}{3c^2}$$

Hence

$$\dot{\gamma}(t,r) = +A\,\dot{\rho}_0(t)\,r^2, \qquad \ddot{\gamma}(t,r) = +A\,\ddot{\rho}_0(t)\,r^2$$

Substituting into $q_{\text{eff}} = 1 - \gamma\,\ddot{\gamma}/\dot{\gamma}^2$ gives

$$\frac{\gamma\,\ddot{\gamma}}{\dot{\gamma}^2} = \gamma(t,r)\,\frac{A\,\ddot{\rho}_0(t)\,r^2}{A^2\,\dot{\rho}_0^2(t)\,r^4} = +\,\frac{\gamma(t,r)}{A\,r^2}\,\frac{\ddot{\rho}_0(t)}{\dot{\rho}_0^2(t)}$$

Therefore

$$q_{\text{eff}}(t,r) = 1 - \frac{\gamma\,\ddot{\gamma}}{\dot{\gamma}^2} = 1 - \frac{\gamma(t,r)}{A\,r^2}\,\frac{\ddot{\rho}_0(t)}{\dot{\rho}_0^2(t)}$$

Using $A = 2\pi G/3c^2$, we can write this as

$$q_{\text{eff}}(t,r) = 1 - \frac{3c^2}{2\pi G}\,\frac{\gamma(t,r)}{r^2}\,\frac{\ddot{\rho}_0(t)}{\dot{\rho}_0^2(t)}$$

which is manifestly dimensionless.

11.7.3 Physical interpretation

The effective deceleration parameter $q_{\text{eff}}(t,r)$ in FSG cosmology, is completely controlled by the sign of $\ddot{\rho}_0(t)$, i.e. by whether the rest–mass density of the Universe is accelerating or decelerating in time.

Case 1 — $\ddot{\rho}_0(t) > 0$ (convex evolution)

$$\frac{\ddot{\rho}_0}{\dot{\rho}_0^2} > 0 \quad \Rightarrow \quad q_{\text{eff}} < 1$$

The second term is positive and reduces q_{eff} below the kinematical value 1. Sufficiently rapid density acceleration may even give

$$q_{\text{eff}} < 0$$

i.e. effective acceleration of the radial flow $r(t)$ without curvature or a cosmological constant.

Case 2 — $\ddot{\rho}_0(t) < 0$ (concave evolution)

$$\frac{\ddot{\rho}_0}{\dot{\rho}_0^2} < 0 \quad \Rightarrow \quad q_{\text{eff}} > 1$$

The second term becomes negative and pushes q_{eff} above 1, corresponding to enhanced effective deceleration.

Case 3 — $\ddot{\rho}_0(t) = 0$ (linear evolution)

$$q_{\text{eff}} = 1$$

This is the neutral kinematical value.

Summary

$$\begin{aligned} q_{\text{eff}} < 1 &\Leftrightarrow \ddot{\rho}_0 > 0 \qquad \text{(effective acceleration)} \\ q_{\text{eff}} > 1 &\Leftrightarrow \ddot{\rho}_0 < 0 \qquad \text{(enhanced deceleration)} \\ q_{\text{eff}} = 1 &\Leftrightarrow \ddot{\rho}_0 = 0 \qquad \text{(linear regime)} \end{aligned}$$

Thus q_{eff} in FSG cosmology is a *purely dynamical* quantity, arising entirely from the temporal behavior of the rest–mass density, and not from curvature or additional geometric degrees of freedom.

11.7.4 Conceptual closure

In Flat Spacetime Gravity the effective deceleration parameter does not arise from a scale factor $a(t)$, a curved metric, or a fluid in an expanding geometry. Instead, it is a directly measurable kinematic diagnostic of how the local gravitational time–rate factor $\gamma(t,r)$ evolves under the influence of the changing matter density $\rho_0(t)$. Cosmic acceleration or deceleration is therefore interpreted as a dynamical property of the matter content and its temporal evolution in a strictly Minkowskian spacetime, rather than as a property of the geometry itself.

11.8 The Apparent Acceleration Equation

11.8.1 Differentiation of the effective velocity

In standard cosmology the cosmic acceleration equation follows from the Einstein–Friedmann system. In Flat Spacetime Gravity, where spacetime remains strictly Minkowskian and all "cosmic evolution" is encoded in the time dependence of the gravitational time–rate factor $\gamma(t,r)$, the corresponding acceleration law must instead be derived from the effective recession kinematics

$$v_{\text{eff}}(t,r) = \frac{1}{\gamma(t,r)} \frac{\partial \gamma(t,r)}{\partial t} r = H_{\text{eff}}(t,r)\, r$$

The effective acceleration is defined kinematically as

$$a_{\text{eff}}(t,r) \equiv \frac{\partial}{\partial t}[H_{\text{eff}}(t,r)\, r]$$

Since r is a fixed comoving label in Flat Spacetime Gravity,

$$a_{\text{eff}}(t,r) = r\,\frac{\partial H_{\text{eff}}(t,r)}{\partial t} = r\,\dot{H}_{\text{eff}}(t,r)$$

Using

$$H_{\text{eff}}(t,r) = \frac{\dot{\gamma}}{\gamma}$$

we obtain

$$\dot{H}_{\text{eff}} = \frac{\ddot{\gamma}}{\gamma} - \frac{\dot{\gamma}^2}{\gamma^2}$$

Thus

$$a_{\text{eff}}(t,r) = r\left(\frac{\ddot{\gamma}}{\gamma} - \frac{\dot{\gamma}^2}{\gamma^2}\right)$$

This is the exact flat-spacetime analogue of the cosmic acceleration equation: it describes the apparent "cosmic acceleration" directly from *the time evolution of the gravitational slowdown factor* γ.

11.8.2 Substitution of the homogeneous FSG cosmology

From previous subsections we have

$$\gamma(t,r) = 1 + A\,\rho_0(t)\,r^2, \qquad A = \frac{2\pi G}{3c^2}$$

Hence

$$\dot{\gamma} = +A\,\dot{\rho}_0(t)\,r^2, \qquad \ddot{\gamma} = +A\,\ddot{\rho}_0(t)\,r^2$$

Substituting into the formula for a_{eff} gives

$$a_{\text{eff}}(t,r) = r\left[\frac{A\,\ddot{\rho}_0 r^2}{1 - A\,\rho_0 r^2} - \frac{(A\,\dot{\rho}_0 r^2)^2}{(1 - A\,\rho_0 r^2)^2}\right]$$

After simplification,

$$a_{\text{eff}}(t,r) = +\frac{A\,r^3\,\ddot{\rho}_0(t)}{\gamma(t,r)} - \frac{A^2\,r^5\,\dot{\rho}_0^2(t)}{\gamma^2(t,r)}$$

Using $A = 2\pi G/3c^2$, we can write this as

$$a_{\text{eff}}(t,r) = +\frac{2\pi G\,r^3\,\ddot{\rho}_0(t)}{3c^2\gamma(t,r)} - \frac{4\pi^2 G^2\,r^5\,\dot{\rho}_0^2(t)}{9c^4\gamma^2(t,r)}$$

11.8.3 Physical interpretation

Three features stand out:

1. *Dependence on density evolution.* The apparent acceleration is driven entirely by the first- and second-time derivatives of the cosmic rest–mass density $\rho_0(t)$. This provides a purely dynamical mechanism for acceleration or deceleration *without geometric expansion.* Spacetime remains flat — only clocks slow down under evolving density.

2. *Two distinct contributions.*

$$a_{\text{eff}} = a_{(\text{I})} + a_{(\text{II})}$$

with

$$a_{(\text{I})} \propto \ddot{\rho}_0(t), \qquad a_{(\text{II})} \propto \dot{\rho}_0^2(t)$$

The first term dominates the sign of the acceleration; the second term acts as a universal positive drift term, reflecting the cumulative weakening of the sourceful gravitational field as density decreases.

3. *The meaning of "acceleration"* in Flat Spacetime Gravity. It does *not* correspond to stretching of spatial distances. Instead, it corresponds to the way recession rates evolve because the local gravitational time-rate factor $\gamma(t,r)$ changes in time. A negative a_{eff} corresponds to "effective cosmic speed-up", a positive a_{eff} to slow-down.

11.8.14 Conceptual closure

The FSG acceleration equation:

$$a_{\mathrm{eff}}(t,r) = r\left(\frac{\ddot{\gamma}}{\gamma} - \frac{\dot{\gamma}^2}{\gamma^2}\right)$$

reveals that cosmic acceleration is determined not by spacetime curvature or a geometric scale factor but by the dynamical competition between $\ddot{\gamma}$ and $\dot{\gamma}^2$ — i.e., by the evolution of the matter density in a fixed Minkowskian background. This provides a clean, geometry–free alternative to the usual GR interpretation of cosmic acceleration.

11.9 Friedmann–Type Evolution Law

11.9.1 Density evolution as the driver of cosmic kinematics

With the effective acceleration law derived in this section,

$$a_{\mathrm{eff}}(t,r) = r\left(\frac{\ddot{\gamma}}{\gamma} - \frac{\dot{\gamma}^2}{\gamma^2}\right)$$

Flat Spacetime Gravity admits a natural analogue of the second Friedmann equation. To proceed, we exploit the homogeneity of the cosmic rest–mass density $\rho_0(t)$ and the exact form of the cosmological time–rate factor:

$$\gamma(t,r) = 1 + A\,\rho_0(t)\,r^2, \qquad A = \frac{2\pi G}{3c^2}$$

Differentiating the above expression yields

$$\dot{\gamma} = +A\,\dot{\rho}_0(t)\,r^2, \qquad \ddot{\gamma} = +A\,\ddot{\rho}_0(t)\,r^2$$

Substituting these into the acceleration expression gives the exact FSG evolution law:

$$\frac{a_{\text{eff}}(t,r)}{r} = \frac{A\,r^2}{\gamma(t,r)}\,\ddot{\rho}_0(t) - \frac{A^2\,r^4}{\gamma^2(t,r)}\,\dot{\rho}_0^2(t)$$

Since $Ar^2 \ll 1$ for all cosmological scales of interest (Section 10.3), the leading behaviour is

$$\frac{a_{\text{eff}}}{r} \simeq A\,\ddot{\rho}_0(t) - A^2 r^2\,\dot{\rho}_0^2(t) + \mathcal{O}(A^3 r^4)$$

11.9.2 A homogeneous limit: the FSG analogue of Friedmann–II

At the homogeneity scale, where the r–dependent correction is negligible, the law reduces to:

$$\frac{a_{\text{eff}}(t)}{r} = A\,\ddot{\rho}_0(t) + \mathcal{O}(A^2\,\dot{\rho}_0^2)$$

with

$$A = \frac{2\pi G}{3c^2}$$

This is the precise FSG counterpart of the "cosmic acceleration equation''. But instead of being geometrical, as in GR, the Flat Spacetime Gravity law is purely *dynamical*: the Universe accelerates or decelerates according to the second time derivative of its rest–mass density.

11.9.3 Interpretation and comparison with standard cosmology

In GR, the acceleration equation is

$$\frac{\ddot{a}}{a} = -\frac{4\pi G}{3}(\rho + 3p/c^2)$$

In Flat Spacetime Gravity we instead find

$$\frac{a_{\text{eff}}}{r} \propto \ddot{\rho}_0(t)$$

so:

- *Geometry plays no role.* There is no scale factor $a(t)$, no curvature term and no stress–energy tensor. All “cosmic acceleration” derives from the dynamical behaviour of the density through the gravitational slowdown factor γ.

- *Acceleration arises even for dust.* In GR, pressure terms are needed for non-zero acceleration. In Flat Spacetime Gravity, even a pure dust density $\rho_0(t)$ can generate acceleration if $\ddot{\rho}_0 \neq 0$.

- *Deceleration and acceleration unified.* A decreasing density ($\dot{\rho}_0 < 0$) automatically gives a positive velocity drift; a concave density function ($\ddot{\rho}_0 < 0$) yields *apparent cosmic acceleration* — entirely without spacetime curvature or dark energy.

11.9.4 Conceptual closure

The effective Friedmann–type law,

$$\frac{a_{\text{eff}}(t)}{r} = +\frac{2\pi G}{3c^2}\,\ddot{\rho}_0(t) \;+\; \text{higher–order density terms}$$

shows unambiguously that Flat Spacetime Gravity recovers the correct qualitative behavior of cosmic acceleration from a flat background with no expanding geometry. The mechanism is encoded entirely in the time-dependent slowdown of clocks, via the evolution of $\gamma(t, r)$, and not in the expansion of space.

11.10 Continuity and Density Evolution

11.10.1 Continuity equation

In a homogeneous dust configuration within Flat–Spacetime Gravity, the matter density $\rho_0(t)$ depends only on the cosmic time t, while each comoving shell at radius r experiences an effective recession velocity

$$v_{\text{eff}}(t,r) = r\,H_{\text{eff}}(t,r)$$

where the FSG Hubble function is defined kinematically by

$$H_{\text{eff}}(t,r) = \frac{1}{\gamma(t,r)}\frac{\partial\gamma(t,r)}{\partial t}$$

$$\gamma(t,r) = 1 - \frac{\phi_{\text{cosm}}(t,r)}{c^2}$$

Mass conservation for a homogeneous dust distribution is encoded in the usual continuity law

$$\dot{\rho}_0(t) + 3\,H_{\text{eff}}(t,R)\,\rho_0(t) = 0$$

where R is an arbitrary reference radius (homogeneity ensures that the choice of R does not affect the form of the equation). Substituting the definition of H_{eff} gives the differential equation

$$\dot{\rho}_0(t) + 3\,\frac{\dot{\gamma}(t,R)}{\gamma(t,R)}\,\rho_0(t) = 0$$

11.10.2 Exact solution for $\rho_0(t)$

We solve this first–order linear equation by separation of variables:

$$\frac{\dot{\rho}_0(t)}{\rho_0(t)} = -3\,\frac{\dot{\gamma}(t,R)}{\gamma(t,R)}$$

Integrating with respect to t,

$$\int_{t_0}^{t}\frac{\dot{\rho}_0(t')}{\rho_0(t')}\,dt' = -3\int_{t_0}^{t}\frac{\dot{\gamma}(t',R)}{\gamma(t',R)}\,dt'$$

Hence

$$\ln[\rho_0(t)] - \ln[\rho_0(t_0)] = -3\{\ln[\gamma(t,R)] - \ln[\gamma(t_0,R)]\}$$

which yields

$$\ln[\rho_0(t)] = \ln[\rho_0(t_0)] - 3\ln[\gamma(t,R)] + 3\ln[\gamma(t_0,R)]$$

Exponentiating both sides we obtain the exact solution

$$\rho_0(t) = \rho_0(t_0)\left[\frac{\gamma(t_0,R)}{\gamma(t,R)}\right]^3$$

Thus, in Flat Spacetime Gravity the homogeneous cosmic density decays (or grows) according to the cube of the inverse time–rate factor $\gamma(t,R)$. No scale factor $a(t)$ or explicit metric expansion is introduced: the entire evolution is driven by the temporal behaviour of the flat–spacetime gravitational slowdown factor γ.

This relation is fully general and exact, independently of any particular functional form of $\gamma(t,R)$.

11.10.3 Summary

The key result of this subsection is the exact law for $\rho_0(t)$ which shows that in Flat Spacetime Gravity the homogeneous density is controlled entirely by the time dependence of the gravitational time–rate factor $\gamma(t,R)$ in a flat spacetime. This provides the natural FSG counterpart of the usual $\rho \propto a^{-3}$ law of FRW cosmology, but expressed purely through γ rather than through a geometric scale factor.

This completes the cosmological core of Flat Spacetime Gravity, demonstrating that cosmic kinematics can be fully reproduced in a Minkowski spacetime without invoking curvature, scale factors, or dark energy.

11.11 Concluding Remarks

The cosmological sector of Flat–Spacetime Gravity, as developed in this section, demonstrates an internally consistent and conceptually transparent picture of large–scale gravitational dynamics without ever departing from the Minkowski metric. The homogeneous matter density $\rho_0(t)$ evolves according to the exact law

$$\rho_0(t) = \rho_0(t_0)\left[\frac{\gamma(t_0,R)}{\gamma(t,R)}\right]^3$$

$$\gamma(t,r) = 1 - \frac{\phi_{\text{cosm}}(t,r)}{c^2}$$

$$\phi_{\text{cosm}}(t,r) = \frac{2\pi G}{3}\rho_0(t)\, r^2$$

showing that cosmic evolution is driven entirely by the temporal behavior of the gravitational time–rate factor $\gamma(t,r)$ rather than by any geometrical expansion of space. This is the FSG analogue of the familiar $\rho \propto a^{-3}$ law of FRW cosmology, but expressed through dynamical time–dilation rather than through a metric scale factor.

The effective acceleration of matter shells is controlled by the radial derivative of the potential,

$$a_{\text{eff}}(t,r) = -\frac{\partial\phi(t,r)}{\partial r} = -\frac{4\pi G}{3}\rho_0(t)\, r$$

and the recession kinematics follow from the definition of the time–rate–based Hubble function

$$H_{\text{eff}}(t,r) = \frac{1}{\gamma(t,r)}\frac{\partial\gamma(t,r)}{\partial t}$$

$$v_{\text{eff}}(t,r) = r\, H_{\text{eff}}(t,r)$$

These relations reveal that in FSG cosmic "expansion" is encoded purely in the evolution of $\gamma(t,r)$, with no geometric stretching of spatial distances. Every observationally relevant quantity (density evolution, effective acceleration, recession velocity) is determined directly from the dynamics of the flat–spacetime gravitational field.

In summary, this section establishes that Flat Spacetime Gravity possesses a fully workable and self–contained cosmological framework: matter conservation, large–scale dynamics, and recession kinematics all emerge from the same underlying principle—the time evolution of the gravitational rate factor γ in a strictly flat spacetime. This stands as a conceptually elegant counterpart to curved–spacetime cosmology, highlighting the internal consistency and structural power of the Flat Spacetime Gravity formulation developed throughout this work.

12. Conclusion

In this work we have developed a complete dynamical framework for gravitation formulated entirely within flat Minkowski spacetime. Without invoking spacetime curvature or a dynamical metric, the theory is built upon antisymmetric field tensors, observer–dependent projections, and a consistent Lagrangian– Hamiltonian dynamics that reproduces the principal phenomenology traditionally attributed to general relativistic geometry.

Starting from the kinematical foundations of four–velocity fields and antisymmetric rotational tensors, a general geometric decomposition into orbital and Fermi components was established. This decomposition provides a transparent separation between spatial dynamics and mixed space–time coupling, allowing the gravitational interaction to be encoded naturally through a Maxwell–type field bivector $G_{\mu\nu}$ and its associated scalar and vector potentials. The resulting dynamical system admits a well–defined Lagrangian formulation, a canonical Hamiltonian structure, and a consistent semiclassical WKB limit leading to the de Broglie wave interpretation.

Within this unified framework, all principal classical tests of gravitation arise directly from flat– spacetime dynamics. Gravitational redshift follows from time–rate variations associated with the scalar potential, orbital precession effects originate from velocity–dependent field couplings, and the full light–deflection angle is obtained through the orbital interaction alone, with the nominal sourceful term being dynamically suppressed in the null limit by relativistic growth of the proper velocity. At no stage is an appeal to spacetime curvature required for these derivations.

At the quantum level, the theory yields a coherent wave description governed by a covariant Hamiltonian operator and a de Broglie field framework paralleling the mathematical structure of relativistic quantum mechanics, while consistently preserving the flat background geometry. In the semiclassical regime, particle trajectories, wave propagation, and classical motion emerge self–consistently as complementary limits of the same underlying field dynamics.

In the cosmological sector, large–scale expansion phenomena can likewise be characterized without geometric curvature. Apparent Hubble expansion, redshift, density evolution, and effective acceleration arise from the dynamical evolution of the gravitational time–rate factor $\gamma(t,r)$ alone, with recession velocities and continuity relations following directly from the flat–spacetime field equations.

Collectively, the results of this work demonstrate that a broad class of gravitational phenomena commonly regarded as intrinsically geometric admit a unified reformulation as dynamical interactions in Minkowski spacetime. The Flat Spacetime Gravity (FSG) framework thus provides a coherent, fully covariant field description in which gravitation, semiclassical wave behavior, quantum structure, and cosmological expansion emerge as interconnected expressions of a single dynamical system, without recourse to spacetime curvature.

Although no claim of formal equivalence with general relativity in all regimes is made or required, the present formulation shows that flat–spacetime dynamics possess sufficient internal structure to capture a wide range of observed gravitational effects. This perspective invites further mathematical refinement, quantum extensions, and targeted phenomenological tests through which the scope and predictive capabilities of the FSG framework may be more fully explored.

Appendix — Glossary of Symbols

Section 2: Kinematical Framework

Symbol	Meaning
c	Speed of light in vacuum
$\eta_{\mu\nu}$	Minkowski metric tensor with signature $(-,+,+,+)$
$\delta^{\mu}_{\ \nu}$	Kronecker delta symbol
x^{μ}	Spacetime coordinates of the test particle
γ	Lorentz parameter, $\gamma = 1/\left(1-\frac{v^2}{c^2}\right)$
$\frac{d}{d\tau}$	Proper–time total derivative along the worldline, $\frac{d}{d\tau} = \gamma\ \frac{d}{dt}$
$\dot{x}^{\mu}$	Proper–time derivatives of the coordinates, $\dot{x}^{\mu} = dx^{\mu}/d\tau$
U^{μ}	Four–velocity of the test body satisfying, $U^{\mu} = \frac{dx^{\mu}}{d\tau}$
$\mathbf{u}$	Proper spatial velocity, $\mathbf{u} = \gamma\mathbf{v}$
α^{μ}	Four–acceleration of the test body, $\alpha^{\mu} = \frac{dU^{\mu}}{d\tau}$
$\mathbf{a}$	Proper spatial acceleration, $\mathbf{a} = \frac{d\mathbf{u}}{d\tau}$
$W_{\mu\nu}$	Kinematical antisymmetric tensor, $W_{\mu\nu} = \partial_{\nu}U_{\mu} - \partial_{\mu}U_{\nu}$
$\boldsymbol{\delta}$	Source field defined via the constitutive relation, $\boldsymbol{\delta} = -\frac{1}{c}W_{0i}$
$\boldsymbol{\xi}$	Vorticity field defined via the spatial components of $W_{\mu\nu}$, $W_{ij} = \varepsilon_{ijk}\xi_k$
$H_{\mu\nu}$	Bivector of specific (per unit mass) angular momentum, $H_{\mu\nu} = X_{\mu}U_{\nu} - X_{\nu}U_{\mu}$
h	Specific orbital angular momentum, $h = \lvert\mathbf{r}\times\mathbf{u}\rvert$
$P(V)_{\mu\nu}$	Spatial projector orthogonal to V^{μ}, $P(V)_{\mu\nu} = \eta_{\mu\nu} + \frac{V_{\mu}V_{\nu}}{c^2}$
$Y_{\mu\nu}$	Unit–mass stress–energy tensor, $Y_{\mu\nu} = U_{\mu}U_{\nu}$

Symbol	Meaning
$\Omega_{\mu\nu}^{(\mathrm{orb})}$	Orbital (spatial) projection of $W_{\mu\nu}$, $\Omega^{(orb)} = P\,\mathrm{W}\,\mathrm{P}$
$\boldsymbol{\omega}_{\mathrm{orb}}$	Orbital projection vector, $\Omega_{jk}^{(\mathrm{orb})} = \varepsilon_{ijk}\omega_{\mathrm{orb,k}}$
$\Omega_{\mu\nu}^{(\mathrm{F})}$	Fermi tensor component associated with relativistic frame rotation, $\Omega_{\mu\nu}^{(\mathrm{F})} = \frac{1}{\mathrm{c}^2}\,\mathrm{U} \wedge \alpha$
$\boldsymbol{\omega}_{\mathrm{F}}$	Fermi precession vector, $\boldsymbol{\omega}_{\mathrm{F}} = \frac{1}{c^2}\,(\mathbf{u} \times \mathbf{a})$
$\Omega_{\mu\nu}^{(\mathrm{F})}$	Thomas tensor component associated with relativistic frame rotation, $\Omega_{\mu\nu}^{(\mathrm{T})} = \frac{1}{\mathrm{c}^2}\,\alpha \wedge U$
$\boldsymbol{\omega}_{\mathrm{T}}$	Thomas precession vector, $\boldsymbol{\omega}_{\mathrm{T}} = \frac{1}{c^2}\,(\mathbf{a} \times \mathbf{u})$

Section 3: Flat-spacetime gravitational field

Symbol	Meaning
Φ^{μ}	Gravitational 4–potential, $\Phi^{\mu} = \left(\frac{\phi}{c}, \mathbf{A}\right)$
ϕ	Scalar gravitational potential
$\mathbf{A}$	Vector gravitational potential
$G_{\mu\nu}$	Gravitational field bivector, $G_{\mu\nu} = \partial_{\mu}\Phi_{\nu} - \partial_{\nu}\Phi_{\mu}$
$^{*}G_{\mu\nu}$	Dual gravitational bivector, $^{*}G_{\mu\nu} = 1/2\,\varepsilon_{\mu\nu\alpha\beta}G^{\alpha\beta}$
$\mathbf{g}$	Sourceful gravitational field, $\mathbf{g} = -\frac{1}{c}G_{0i}$
$\mathbf{b}$	Rotational gravitational field, $G_{ij} = +\,\varepsilon_{ijk}\,b_k$
I_1	First invariant of Gravitational field tensor, $\mathrm{I}_1 = \frac{1}{2}G_{\mu\nu}G^{\mu\nu} = \lvert\mathbf{b}\rvert^2 - \frac{\lvert\mathbf{g}\rvert^2}{c^2}$
I_2	Second invariant of Gravitational field tensor, $\mathrm{I}_2 = \frac{1}{4}\,\varepsilon^{\mu\nu\alpha\beta}G_{\mu\nu}G_{\alpha\beta} = -\frac{1}{\mathrm{c}}\,\mathbf{g}\cdot\mathbf{b}$
V_0	Rest volume of the test body
m_0	Rest mass of the test body

Symbol	Meaning
ρ_0	Rest–mass density of gravitational sources, $\rho_0 = \frac{dm_0}{dV_0}$
j^μ	Four–current of gravitational sources, $j^\mu = \rho_0 U^\mu$
$\mathbf{j}$	Gravitational mass current density, $\mathbf{j} = \rho_0 \mathbf{u}$
∇	Spatial gradient operator
Grad	Four-dimensional grad operator, Grad $= \partial_\mu$
Div	Four-dimensional div operator, Div $= \partial_\mu \cdot$
Curl	Four-dimensional curl operator, Curl $= \partial_\mu \wedge$
$\Box$	Covariant d'Alembert operator, $\Box = \partial_\mu \partial^\mu = \nabla^2 - 1/c^2\, \partial_t^2$
k^μ	The wave four-vector, $k^\mu = (\omega/c, \mathbf{k})$
u_{field}	Energy density, $u_{\text{field}} = \frac{c^2}{8\pi G}\left(\frac{\lvert\mathbf{g}\rvert^2}{c^2} + \lvert\mathbf{b}\rvert^2\right)$
$\mathbf{S}$	Poynting vector, $\mathbf{S} = \frac{c^2}{4\pi G}\, \mathbf{g} \times \mathbf{b}$
$T^{\mu\nu}_{\text{field}}$	Energy–momentum tensor, $T^{\mu\nu}_{\text{field}} = \frac{c^2}{4\pi G}\left(G^\mu{}_\rho\, G^{\nu\rho} - \frac{1}{4}\eta^{\mu\nu} G_{\alpha\beta} G^{\alpha\beta}\right)$
$T^{\mu\nu}_{\text{matter}}$	Matter Energy–Momentum Tensor, $T^{\mu\nu}_{\text{matter}} = \rho_0\, U^\mu U^\nu$
$\mathbf{J}$	Mass-current dipole moment, $\mathbf{J} = \int \mathbf{x}' \times \mathbf{j}(\mathbf{x}')\, d^3x'$
$\mathbf{S}_{\text{spin}}$	Spin (or internal angular momentum) of the test body
$\boldsymbol{\tau}$	Gravitational torque, $\boldsymbol{\tau} = \frac{G}{c^2} \mathbf{S}_{\text{spin}} \times \mathbf{b}$

Section 4: Effective Potential

Symbol	Meaning
N^μ	Four-velocity of the source, $N^\mu = (c, 0,0,0)$
$\widehat{H}_{\mu\nu}$	Projected gravitational rotational tensor, $\widehat{H}^{\mu\nu} = P^\mu{}_\alpha\, H^{\alpha\beta}\, P_\beta{}^\nu$

Symbol	Meaning
$\hat{\mathbf{h}}$	Spatial dual vector of $\hat{H}_{ij}$, $\hat{H}_{ij} = \varepsilon_{ijk}\hat{h}_k$
$\Phi_\mu^{(\mathrm{orb})}$	Orbital our–potential, $\Phi^{(\mathrm{orb})\mu} = \left(\frac{\phi^{(\mathrm{orb})}}{c}, \mathbf{A}^{(\mathrm{orb})}\right) = \frac{1}{c^2} G^{\mu\rho} \hat{H}_{\rho\nu} U^\nu$
$\phi^{(\mathrm{orb})}$	Orbital Scalar potential, $\phi^{(\mathrm{orb})} = \frac{1}{c^2} \mathbf{g} \cdot (\mathbf{u} \times \mathbf{h})$
$\mathbf{A}^{(\mathrm{orb})}$	Orbital Vector potential, $A^{(\mathrm{orb})i} = \frac{1}{c^2} G^{i\rho} \hat{H}_{\rho\nu} U^\nu$
${G^{(\mathrm{orb})}}_{\mu\nu}$	Orbital gravitational field bivector ${G^{(\mathrm{orb})}}_{\mu\nu} = \partial_\mu {\Phi^{(\mathrm{orb})}}_\nu - \partial_\nu {\Phi^{(\mathrm{orb})}}_\mu$
$\mathbf{g}^{(\mathrm{orb})}$	Orbital sourceful gravitational field, $\mathbf{g}^{(\mathrm{orb})} = -\frac{1}{c} {G^{(\mathrm{orb})}}_{0i}$
$\mathbf{b}^{(\mathrm{orb})}$	Orbital rotational gravitational field, ${G^{(\mathrm{orb})}}_{ij} = +\varepsilon_{ijk} {b^{(orb)}}_k$
$\Phi_\mu^{(\mathrm{eff})}$	Effective gravitational four–potential, $\Phi^{(\mathrm{eff})\mu} = \Phi^\mu + \Phi^{(\mathrm{orb})\mu}$
$\phi^{(\mathrm{eff})}$	Scalar effective potential, $\phi^{(\mathrm{eff})} = \phi + \phi^{(\mathrm{orb})}$
$\mathbf{A}^{(\mathrm{eff})}$	Vector effective potential, $\mathbf{A}^{(\mathrm{eff})} = \mathbf{A} + \mathbf{A}^{(\mathrm{orb})}$
${G^{(\mathrm{eff})}}_{\mu\nu}$	Effective gravitational field bivector ${G^{(\mathrm{eff})}}_{\mu\nu} = \partial_\mu {\Phi^{(\mathrm{eff})}}_\nu - \partial_\nu {\Phi^{(\mathrm{eff})}}_\mu$
$\mathbf{g}^{(\mathrm{eff})}$	Effective sourceful gravitational field, $\mathbf{g}^{(\mathrm{eff})} = \mathbf{g} + \mathbf{g}^{(\mathrm{orb})}$
$\mathbf{b}^{(\mathrm{eff})}$	Effective rotational gravitational field, $\mathbf{b}^{(\mathrm{eff})} = \mathbf{b} + \mathbf{b}^{(\mathrm{orb})}$
E	Conserved energy per unit mass of the test body, $E = \gamma c^2 - c^2 + \phi^{(\mathrm{eff})}$
$\Pi_\mu^{(\mathrm{eff})}$	Generalized momentum per unit mass, $\Pi_\mu^{(\mathrm{eff})} = \left(\mathrm{U}_\mu + {\Phi^{(\mathrm{eff})}}_\mu\right)$
$\mathrm{P}_\mu^{(\mathrm{eff})}$	Generalized momentum, $\mathrm{P}_\mu^{(\mathrm{eff})} = \mathrm{m}_0\left(\mathrm{U}_\mu + {\Phi^{(\mathrm{eff})}}_\mu\right)$
$\mathbf{P}_{\mathrm{NR}}$	Newtonian spatial part of generalized momentum, $\mathbf{P}_{\mathrm{NR}} = m_0\mathbf{v} + m_0\mathbf{A}^{(\mathrm{eff})}$

Section 5: Astronomical Tests

Symbol	Meaning
$u(\theta)$	Standard Binet variable, $u(\theta) = \frac{1}{r(\theta)}$
$\Delta\theta$	Perihelion advance angle, $\Delta\theta = \frac{6\pi GM}{a(1-e^2)c^2}$
b	Impact parameter of the scattering trajectory
h	Specific orbital angular momentum, $h = b\,\lvert\mathbf{u}\rvert$
$\Delta\theta_S$	Sourceful contribution to the deflection angle, $\Delta\theta_S = \frac{2GM}{b\,u^2}$
$\Delta\theta_{\text{orb}}$	Orbital contribution to the deflection angle, $\Delta\theta_{\text{orb}} = \frac{4GM\,h^2}{c^2\,u^2\,b^3}$
$\Delta\theta$	Total gravitational deflection angle of light or massive test particles, $\Delta\theta = \Delta\theta_S + \Delta\theta_{\text{orb}}$
$\Delta\theta_{\text{light}}$	Gravitational deflection angle of light, $\Delta\theta_{\text{light}} = \frac{4GM\,h^2}{c^2\,u^2\,b^3}$
ν_0	Emitted radiation frequency
ν	Observed radiation frequency at the reception point r, $\nu = \left.\frac{d\tau}{dt}\right\rvert_r$
$\Delta t_{\text{Shapiro}}$	Gravitational time delay (Shapiro delay), $\Delta t_{\text{Shapiro}} \simeq \frac{4GM}{c^3}\ln\left(\frac{4r_1r_2}{b^2}\right)$

Section 6: Relativistic Lagrangian Formulation

Symbol	Meaning
L	Relativistic gravitational Lagrangian
L_{free}	Free–particle Lagrangian term, $L_{\text{free}} = m_0\Phi^{(\text{eff})}{}_{\mu}\,U^{\mu}$
L_{int}	Interaction Lagrangian term, $L_{\text{int}} = -m_0c^2$
L_{part}	Particle Lagrangian, $L_{\text{part}} = L_{\text{free}} + L_{\text{int}}$

Symbol	Meaning
L_{field}	Field Lagrangian, $= -\frac{c^2}{16\pi G} G_{\mu\nu}G^{\mu\nu}$
L_{tot}	Total Lagrangian, $L_{\text{tot}} = L_{\text{part}} + L_{\text{field}}$
S	Action functional, $S = \int L\, d\tau$
S_{free}	Free–particle Action term, $S_{\text{free}} = \int L_{\text{free}}\, d\tau$
S_{int}	Interaction Action term, $S_{\text{int}} = \int L_{\text{int}}\, d\tau$
S_{part}	Particle Action, $S_{\text{part}} = S_{\text{free}} + S_{\text{int}}$
S_{field}	Field Action, $S_{\text{field}} = \int L_{\text{field}}\, d^4x$
S_{tot}	Total Action, $S_{\text{tot}} = S_{\text{part}} + S_{\text{field}}$
H	Hamiltonian, $H = \mathbf{v} \cdot \frac{\partial \mathcal{L}}{\partial \mathbf{v}} - \mathcal{L}$

Section 7: Relativistic Hamiltonian Formulation

Symbol	Meaning
$\mathbf{p}$	Spatial canonical momentum components, $\mathbf{p} = \frac{\partial \mathcal{L}}{\partial \mathbf{v}}$
$\mathbf{\Pi}$	Mechanical momentum, $\mathbf{\Pi} = \mathbf{p} - m_0\, \mathbf{A}^{(\text{eff})} = m_0\, \mathbf{u}$
H	Hamiltonian operator, $H = \mathbf{p} \cdot \mathbf{v} - \mathcal{L}$
(q_i, p_i)	Canonical coordinate–momentum pairs in phase space, $\dot{q}^i = \frac{\partial H}{\partial p_i}, \dot{p}_i = -\frac{\partial H}{\partial q^i}$
$S(\mathbf{x}, t)$	Hamilton–Jacobi principal function, $\mathbf{p} = \nabla S$
Π^μ	Canonical momentum conjugate to Φ_μ, $\Pi^\mu = \frac{\partial \mathcal{L}_{\text{field}}}{\partial(\partial_t \Phi_\mu)}$
$\hbar$	Reduced Planck constant, $\hbar = \frac{h}{2\pi}$
$\psi(x)$	Scalar wavefunction, $\psi(x) = A(x)\, \exp\left(\frac{i}{\hbar} S(x)\right)$
ℓ	Angular momentum, $\ell = m_0 h$
$\oint p_i\, dq_i$	Action integral defining the semiclassical quantisation condition

Symbol	Meaning
J_ϕ	Angle canonical action, $J_\phi = \frac{1}{2\pi}\oint p_\phi \, d\phi$
J_r	Radial canonical action, $J_r = \frac{1}{2\pi}\oint p_r \, dr$
ω_ϕ	Angle frequency, $\omega_\phi = \frac{\partial H}{\partial J_\phi}$
ω_r	Radial frequency, $\omega_r = \frac{\partial H}{\partial J_r}$
$\hat{p}_\mu$	Canonical momentum operator, $\hat{p}_\mu = -i\hbar \, \partial_\mu$
$\hat{\Pi}_\mu$	Mechanical momentum operator, $\hat{\Pi}_\mu = \hat{p}_\mu - m_0 \Phi_\mu^{(\text{eff})}$
Ψ	Quantum wavefunction of the test particle, $\psi = e^{-iHt/\hbar} \chi(\mathbf{x})$

Section 8: De Broglie–Gravitation Connection

Symbol	Meaning
$d\Sigma_{\mu\nu}$	Hyper-surface element, $d\Sigma_{\mu\nu} = \frac{1}{2} \epsilon_{\mu\nu\alpha\beta} \, dx^\alpha \wedge dx^\beta$
P^μ	Four-momentum of the test particle, $P^\mu = m_0 U^\mu$
K^μ	Matter wave four–vector defined by $P^\mu = \hbar \, K^\mu$
φ	Phase of the matter wave, $\varphi = K_\mu X^\mu$

Section 9: WKB Structure in FSG and Matter Waves

Symbol	Meaning
α	Maslov index
n_i	Quantum numbers defined through WKB quantisation conditions, $I_i = n_i \hbar$
$\mathcal{H}$	Hamiltonian density, $\mathcal{H} = \frac{2\pi G}{c^2} \Pi^i \Pi_i + \frac{c^2}{8\pi G} (\nabla \times \mathbf{\Phi})^2$

Symbol	Meaning
H_{int}	Interaction vertex, $H_{\text{int}} \sim m_0\left[a_{\mathbf{k},\lambda}\, e^{-ik\cdot x(t)} + a^{\dagger}_{\mathbf{k},\lambda}\, e^{+ik\cdot x(t)}\right](\gamma\, \epsilon_0 - \mathbf{u}\cdot\boldsymbol{\epsilon})$
$\mathcal{A}_{i\to f}$	Transition amplitude, $\mathcal{A}_{i\to f} = -\frac{i}{\hbar}\int_{-\infty}^{+\infty} d\,t\ \langle f \mid H_{\text{int}}(t) \mid i\rangle$
$\mathcal{M}(k,\lambda)$	Basic vertex, $\mathcal{M}(k,\lambda) \sim m_0 \int dt\ U^{\mu}(t)\, \epsilon^{*}_{\mu}(\mathbf{k},\lambda)\, e^{ik\cdot x(t)}$
$D_{\mu\nu}(x - x')$	Propagator, $D_{\mu\nu}(x - x') = \eta_{\mu\nu}\, \delta^4(x - x')$
$D_{ij}(\mathbf{k},\omega)$	Spatial propagator, $D_{ij}(\mathbf{k},\omega) = \frac{1}{\omega^2/c^2 - \lvert\mathbf{k}\rvert^2 + i0^+}\left(\delta_{ij} - \frac{k_i k_j}{\lvert\mathbf{k}\rvert^2}\right)$
$\mathcal{A}_{\mathbf{k},\lambda}$	Amplitude for emission, $\mathcal{A}_{\mathbf{k},\lambda} = -\frac{i m_0}{\hbar}\int_{-\infty}^{+\infty} dt\ U^{\mu}(t)\, \epsilon^{*}_{\mu}(\mathbf{k},\lambda)\, e^{ik\cdot x(t)}$
P	Total radiated power, $P = \frac{m_0^2 G}{\pi c^2}\sum_{\lambda=1}^{2}\int d^3k\, \lvert\mathbf{u}_{\perp}(\omega_{\mathbf{k}})\rvert^2 \delta(\omega_{\mathbf{k}} - n\Omega)$

Section 10: Quantum Structure of Flat Spacetime Gravity

Symbol	Meaning
$V_{\text{eff}}(r)$	Effective potential energy, $V_{\text{eff}}(r) = m_0 \phi^{(\text{eff})}(r)$
E_{orb}	Orbital energy, $E_{\text{orb}} = \frac{1}{2} m_0 \left(\frac{dr}{dt}\right)^2 + V_{\text{eff}}(r)$
$p_r(r)$	Radial momentum, $p_r(r) = m_0\, \gamma(r)\, \frac{dr}{dt}$
$E_{n\ell}$	Discrete energy levels, $E_{n\ell} = E_{\text{orb}}(n,\ell)$
Q_{ij}	Quadrupole moment, $Q_{ij} = \int \rho(\mathbf{x}) \left(x_i x_j - \frac{1}{3}\, \delta_{ij}\, \mathbf{x}^2\right) d^3x$

Section 11: Cosmology in Flat Spacetime Gravity

Symbol	Meaning
$\phi_{\text{cosm}}(t,r)$	Cosmological potential, $\phi_{\text{cosm}}(t,r) = -\frac{2\pi G}{3}\rho_0(t)\,r^2 + C(t)$
z	Cosmological redshift factor, $z = \Delta\phi_{\text{cosm}}/c^2$
$v_{\text{eff}}(t,r)$	Effective recession velocity, $v_{\text{eff}}(t,r) = r\frac{d}{dt}(\ln\gamma(t,r))$
$a_{\text{eff}}(t,r)$	Effective acceleration, $a_{\text{eff}}(t,r) = r\frac{d^2}{dt^2}(\ln\gamma(t,r))$
$\mathrm{H}_{\text{eff}}(t,r)$	Effective Hubble factor, $\mathrm{H}_{\text{eff}}(t,r) = \frac{v_{\text{eff}}(t,r)}{r} = \dot{\gamma}/\gamma$
$q_{\text{eff}}(t,r)$	Effective deceleration parameter, $q_{\text{eff}}(t,r) = -\frac{1}{H_{\text{eff}}^2(t,r)}\frac{\partial H_{\text{eff}}(t,r)}{\partial t}$

REFERENCES

[1] A. Einstein, Die Feldgleichungen der Gravitation, Sitzungsberichte der Preussischen Akademie der Wissenschaften, 1915.

[2] A. Einstein, The Meaning of Relativity, Princeton University Press, 1922.

[3] C. Møller, The Theory of Relativity, Oxford University Press, 1952.

[4] L.D. Landau and E.M. Lifshitz, The Classical Theory of Fields, Pergamon Press, 1975.

[5] R.M. Wald, General Relativity, University of Chicago Press, 1984.

[6] C.W. Misner, K.S. Thorne, and J.A. Wheeler, Gravitation, W.H. Freeman, 1973.

[7] O. Heaviside, "A Gravitational and Electromagnetic Analogy", The Electrician, 1893.

[8] I. Ciufolini and J.A. Wheeler, Gravitation and Inertia, Princeton University Press, 1995.

[9] B. Mashhoon, "Gravitoelectromagnetism", Lect. Notes Phys., 562 (2001).

[10] M. Fierz and W. Pauli, "On relativistic field equations for particles of arbitrary spin", Proc. Roy. Soc. A, 1939.

[11] S.N. Gupta, "Gravitation and electromagnetism", Phys. Rev., 96, 1683 (1954).

[12] S. Weinberg, "Photons and Gravitons in S-Matrix Theory", Phys. Rev., 138 B988 (1965).

[13] A.A. Logunov, Relativistic Theory of Gravitation, Nova Science, 1998.

[14] M.V. Berry and K.E. Mount, "Semiclassical approximations in wave mechanics", Rep. Prog. Phys., 1972.

[15] V.P. Maslov and M.V. Fedoriuk, Semiclassical Approximation in Quantum Mechanics, Reidel, 1981.

[16] L. Brillouin, Wave Propagation and Group Velocity, Academic Press, 1960.

[17] L. de Broglie, "Recherches sur la théorie des quanta", Annales de Physique, 1924.

[18] D. Bohm, "A Suggested Interpretation of the Quantum Theory", Phys. Rev., 1952.

[19] P.R. Holland, The Quantum Theory of Motion, Cambridge Univ. Press, 1993.

[20] P.C. Peters & J. Mathews, "Gravitational Radiation from Point Masses", Phys. Rev., 131 (1963).

[21] S. Weinberg, Gravitation and Cosmology, Wiley, 1972.

[22] V.I. Arnold, Mathematical Methods of Classical Mechanics, Springer, 1978.

[23] H. Goldstein, Classical Mechanics, Addison-Wesley, 1980.

[24] J.D. Jackson, Classical Electrodynamics, Wiley, 1998.

[25] R.P. Feynman, Quantum Theory of Gravitation, Lect. Notes, Caltech, 1963.

[26] S. Carlip, "Quantum Gravity: A Progress Report", Rep. Prog. Phys., 2001.

[27] Y. Nambu, "The Use of the Action Principle in Quantum Field Theory", Prog. Theor. Phys., 1950.

[28] V. Mukhanov, Physical Foundations of Cosmology, Cambridge Univ. Press, 2005.

[29] P.A.M. Dirac, The Principles of Quantum Mechanics, Oxford Univ. Press, 1958.

www.ingramcontent.com/pod-product-compliance
Lightning Source LLC
Chambersburg PA
CBHW050523240926
51846CB00003B/675